Backyard Beekeeping

Backyard Beekeeping

C.N. SMITHERS

Kangaroo Press

In memory of the late Don Callaghan,
who introduced me to bees

Acknowledgments

I would like to thank the members of the North Shore Branch of the Amateur Beekeepers' Association who have shared their beekeeping experiences with me and taught me about bees. My wife, Smila, and sons Graeme and Hartley, have helped in many ways with my beekeeping. Bruce White kindly read and criticised the manuscript and made many valuable comments. My wife typed the manuscript and prepared the index, and the photographs were taken by Vanessa Peters and Mark Porter.

Cover design by Darian Causby using slides by Vanessa Peters and Jan Taylor

First published in 1987
Second edition published in 1992
Reprint of second edition published in 1995 by Kangaroo Press Pty Ltd
3 Whitehall Road Kenthurst NSW 2156 Australia
(PO Box 6125 Dural Delivery Centre NSW 2158)
Typeset by G.T. Setters Pty Limited
Printed in Hong Kong by Colorcraft Ltd

ISBN 0 86417 458 6

Contents

Preface

This book is written for those who want to keep a hive or two of bees, to have an interesting, inexpensive hobby which, while being basically fairly simple, challenges ingenuity and at the same time provides the satisfying reward of a little honey for family and friends. There is always something special about eating honey from your own hives.

Many people are worried about being near a beehive. With a little experience and knowledge, confidence will replace fear of bees buzzing around you, and the pleasure of opening hives and handling the bees, watching their progress and activities, and seeing the future honey crop filling the combs, will be well worth the effort and occasional sting. The bee is not a malicious insect. If it is handled gently, with consideration for its comfort and welfare, stings will be few and far between.

To get the most satisfaction from having a few hives you should join a local beekeepers' association. Here you will find the skills, experience and knowledge of long-time beekeepers and you will make new friends. Beekeepers form a fraternity with a spirit of friendliness born of common interest such as has gone from many aspects of modern living. It is refreshing and reassuring to find that it still exists among beekeepers. Beekeeping is a quiet hobby, best done in a leisurely way. It is a hobby which takes time, a seemingly disappearing commodity, and brings relaxation and peace of mind. If you can make it a family hobby, so much the better.

Beekeeping is essentially practical and no amount of reading can replace practical experience. *It is a hobby to enjoy—so enjoy it*. If this book helps to introduce you to bees and sets you on the way to enjoying them it will have fulfilled its purpose.

CHAPTER ONE

Introduction to Beekeeping

The importance of bees

The exploitation of bees by humans began before written history, when wild colonies were robbed and probably largely destroyed in the process. Honey has always been a major source of sugar. In time, after the initial simple management of hives in hollow containers, through the stage of keeping bees in basket-like hives, the removable frame hive was developed. Until the invention of this the comb always had to be destroyed to obtain the honey. This is most wasteful as the time and effort that the bees have to put into remaking comb is better spent in gathering nectar. The history of beekeeping is interesting in itself as an example of progress in human endeavour.

The development of the familiar modern rectangular beehive, consisting of boxes containing removable frames of fairly precise dimensions, has enabled beekeeping and honey production to become a major industry in many countries. Russia, the United States, Mexico, Canada, Argentina and Australia are the world's main honey producers.

Bees are not only important as makers of honey and wax, without them many of the major crops which need pollination by insects could not be grown on the scale necessary to feed us. Many fodder and fruit crops are dependent on bees for pollination for fruit and seed set. Thousands of tonnes of honey are produced each year with Australia being a major exporting country. It has been estimated that the value to agriculture of bees as pollinators of plants is at least 40 times as great as the value of honey as an export, but the return is not obvious because it is not directly measurable in money.

Kinds of honey bee

All bees which are kept or cultured for honey production in Australia belong to one species, the honey bee, which has the scientific name of *Apis mellifera*. In this book "bee" means this species only. This species has a very wide natural distribution in Europe, Asia and Africa and during its long evolutionary history there developed recognisable populations or subspecies in several parts of the range. These subspecies are usually referred to as "races" by beekeepers. Although they evolved in separate parts of the world and have clear characteristics, the races have not evolved so far from each other as to prevent interbreeding. Thus, all races can be crossed with each other.

Three races have been used in Australian beekeeping, the Italian (*Apis mellifera ligustica*), the Carniolan (*Apis mellifera carnica*) and the Caucasian (*Apis mellifera caucasica*). In other parts of the world other subspecies are sometimes used, for example, in South Africa *Apis mellifera adansonii*, the native race, has been domesticated. With the development of beekeeping and the search for improved stocks there has been considerable selection to give "strains" within each race which have specific, desirable characters. Bee breeding is a highly developed science and it is possible to obtain stocks with known pedigrees and very specific characteristics.

Some races are now difficult or impossible to find in the wild because of the deliberate crossing of races in the attempts to obtain better bees and the introduction of bees from one part of the world to another. For example, it is believed that the original European race *(Apis mellifera mellifera)* cannot now be found in a pure form except at Tarraleah, in Tasmania, to which it was taken in early days and where it has not been in contact with other races and hence is still typical of the original race.

Each race has not only physical characteristics of its own, such as size and colour, but also characteristics of behaviour peculiar to it. Just as colour varies, so do the other characteristics and now, one characteristic alone, such as colour or size, does not necessarily imply that the other racial characteristics will also be certain to occur with it. Because of accidental or deliberate interbreeding of races, there is variation in the "races" now found in Australia.

Some general comments can be made but the potential variability of most stocks should always be borne in mind.

The *Italian* race is that most often used in Australia. In typical worker specimens, Italian bees can be recognised by the two to five yellow abdominal bands. They have fairly long tongues and tend to start activities early in spring. Colour and temperament are both very variable but there are many docile "strains" which have been selectively bred and which have reliably predictable characteristics and behaviour. It is the only yellow race in Australia.

Carniolan workers have no yellow abdominal bands but have brown bands of hair. Their hair is short and the overall appearance is of a greyish bee. The tongue is long and activity starts early in spring. Some strains have a strong tendency to swarm.

The *Caucasians*, like the Carniolans, have brown bands but have an overall darker appearance. They have longer tongues than either the Italians or Carniolans. They are advantaged in this way as they can reach deeper into longer flowers which may not be utilised by the other races. They are usually very docile and colonies develop rapidly, although there is a tendency for overwintering colonies to be small. Although they are hard workers they sometimes tend to make the hive a little more difficult to manage because of their excessive use of propolis (vegetable material) to seal up holes and fill gaps. They also have a reputation for being robbers of other hives.

To manage bees it is essential to know a little about the way they live. Beekeepers cannot change that way but can increase their productivity by understanding it and giving them every help. Successful beekeepers help their bees to live well the way they have always done; unsuccessful beekeepers are those who do things contrary to the bees' natural way of life. It is best to begin by learning a little about the bees themselves because success depends on understanding them.

CHAPTER TWO
The Life of the Bee

The three castes

A colony of bees usually consists of individuals of three kinds (castes). Each caste has very definite tasks to carry out and each bee has bodily features which enable it to do this efficiently.

The *queen bee*, usually only one in the colony, has the task of laying all the eggs required to maintain the numbers in the colony. She is capable of laying 1000–2000 eggs a day if required. She is the biggest bee in the colony, being about half as long again as the workers. Her wings do not reach the end of her abdomen and her body tapers somewhat towards the back end. Her tongue is shorter than that of the workers and her sting, which, as in all insects, is primarily an egg-laying instrument, is not barbed. She usually only uses it to kill other queen bees. She does not have special structures on her legs for gathering pollen and she does not have waxproducing glands on the underside of her abdomen.

The *worker bees* are also all females. Their reproductive organs are not fully developed so that they do not, in normal circumstances, lay eggs. The workers are by far the most numerous caste in the colony and are the "bees" familiar to most people. They are those seen visiting flowers. Workers are the smallest of the three castes and are responsible for most of the work carried out in the colony and hence for its day to day wellbeing. The worker is an active flier and her wings reach beyond the end of her abdomen. Her tongue is long and she has a sharp, barbed sting which is used in defence of the colony. She has special combs on the basal part of the tarsus of her hind legs with which she removes pollen caught up on her densely hairy body. She deposits the pollen into special baskets (corbicula) on her back legs. The underside of her abdomen has special glands which produce wax used in making the structure of the honey comb.

The *drones* are the male bees. The drone is more heavily built than the worker and his abdomen is rather square-ended. His wings reach beyond the end of his body. His eyes are very large and almost meet on top of his head. His only function is to fertilise the virgin queens. As males have no egg-laying instrument, the drones cannot sting. He has no pollen-collecting apparatus on his back legs and he does not produce wax.

The three bee castes: queen, worker and drone.

The life cycle

All the bees in a hive pass through the same series of stages during their growth and development, irrespective of caste.

Each starts life as a small egg laid by the queen in one of the cells of the wax comb. She lays only one egg in each cell and from the egg hatches a small, whitish, legless grub (larva). The larva is fed by the worker bees and grows at a very rapid rate. In a few days it is big enough to fill the whole cell. During growth the larva sheds its skin (moults) four times to accommodate the increase in size. The worker bees seal over the mouth of the cell with wax when the larva is fully grown and the larva spins a thin, silken cocoon inside the cell. The larva later moults again, this time to reveal the chrysalis (pupa) which is an immobile, non-feeding stage. The body of the adult develops inside the pupa and eventually emerges by shedding the pupal skin. The bee is then ready to leave its cell and join the rest of the colony. The immature population of larvae and pupae in the hive are referred to as the "brood".

The queen bee is able to determine whether the egg she lays is fertilised or not. A non-fertilised egg develops into a drone. A fertilised egg develops into a female, that is, either a queen bee or a worker. Whether it becomes worker or queen is determined by the type of food fed to the larva by the workers. All larvae are given the same food by the workers for the first two days after hatching, predominantly a substance called royal jelly. This is a very rich product of the pharyngeal glands of the young

workers. They secrete quantities of royal jelly and place it in the cells so that the young larvae literally lie in their food. If a larva is destined to become a queen she is fed mainly on royal jelly throughout her larval period. Other larvae, that is, males, or those destined to become workers, are fed on a mixture of pollen and honey after the second day. Although each individual passes through the same stages during development, the length of time of each differs according to caste. The approximate duration for each stage (in days) is given in the following table.

	Length of stages (in days) for each caste		
	Queen	*Worker*	*Drone*
Egg	3	3	3
Larval feeding	5	6	6
Cell closed on	8th day	9th day	10th day
Spinning cocoon	1	1	3
Non-feeding larva	1	1	2
Pupa	6	10	10
Total	**16**	**21**	**24**

Although the queen is larger than the workers and might be expected to take longer to develop, her developmental period is actually shorter owing to the richness of her diet throughout her larval period.

The young queen will emerge from her cell on about the sixteenth day and will remain in the hive for a few days, wandering over the frames and feeding herself. She has little communication with other bees but will attack another queen if she meets one. Over the next three or four days she leaves the hive to take a few orientation flights through which she becomes familiar with the hive and its surroundings. At some time from about the fifth to the ninth day after emerging from her cell she sets off on her mating flight, although this event can be delayed for several days if necessary. During this flight she will mate with a drone after which she will return to her home hive. Two to four days after mating she is ready to start laying eggs. Normally the queen will not leave the hive again unless the colony absconds or she leaves with a swarm to establish a colony at a new site.

If, for some reason, such as wing deformity, she is not able to leave a hive to mate, the young queen may nevertheless begin to lay eggs. As the eggs will not be fertilised they will all produce drones. Drone-laying queens can usually be detected by the excessively domed caps of the closed cells and by the irregular laying pattern.

The queen can live as long as five years or more

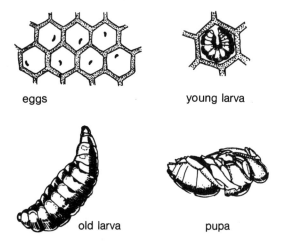

eggs   young larva

old larva   pupa

Like other insects, the bee passes through a number of distinct stages before reaching maturity.

although she may not be at peak productivity after three years.

The worker bee does not leave the hive for the first two weeks or so after emerging from the pupa. She is then known as a "house bee" or "nurse bee" and is responsible for a variety of activities within the hive. After this period she leaves to take flights outside the hive and becomes a "field bee". The length of her life as an adult bee depends very largely on her activity during the time she is a field bee. When there is a lot of nectar available to be collected and the colony is thriving, with a lot of larvae to be provided for, the field bees work very hard, use more energy and live for a shorter period; perhaps for as little as seven weeks. If things are generally slower in the hive, with few larvae to feed, her activity may be reduced and her length of life increased. During winter the bees may live six months or longer if the colony has ample food in store and the workers do not have to maintain a high level of activity to keep the hive warm.

Sometimes some of the workers are able to lay eggs. These, not being fertilised, will produce drones. This usually happens when a colony has lost its queen and has not been able to rear another. The eggs are laid in a very irregular pattern, as are those of a drone-laying queen, and often several are laid in one cell. Probably the best procedure for the beekeeper in such a case is to unite the colony with a healthy one (see "Uniting Colonies").

The drone's only function in life is to fertilise a virgin queen. From about a week after emerging from his cell the drone will take a few short flights outside the hive. After about a fortnight he is sufficiently mature to be able to mate. This he does on a mating flight during which several to many drones compete for the privilege of mating with a queen. During mating his reproductive organs become held firmly in the genital tract of the queen and when the two separate his reproductive organs are torn from his body and remain attached for some time to the queen, even after she has returned to her hive. The drone sooner or later dies from the injury.

The life span of the drones which are not killed through mating depends largely on the state of the colony in which they have been reared. A thriving, healthy colony will tolerate many drones, the workers feeding them because they do not feed themselves. If the colony is under stress, such as from food shortage or low temperatures, the drones are forced from the hive and are not allowed to return or are killed. Even if they are not killed by the workers they cannot fend for themselves outside the hive and soon die of starvation.

Colony size

The number of bees in a colony varies widely depending on season, vigour of the queen, working quality of the workers, availability of food and other factors. Numbers can vary from a few thousands during winter to as many as 100,000 at a time of great vigour and prosperity. In addition to the adult bees there will be thousands of larvae and eggs. For a colony of 50,000 adults there may be as many as 9000 larvae and 6000 eggs as well as another 20,000 as larvae in sealed cells or as pupae. When the weather is warm and at a time when nectar is plentiful, a count of 200 a minute entering the hive would indicate that about 4000 bees are active away from the hive and that the total number of field bees would be in the region of four times as much, that is, about 16,000. There would probably also be about the same number of nurse bees, giving a colony of about 30,000. These figures are, of course, only rough guides as the proportions would vary depending on whether the prosperity of the colony was increasing or declining and its requirements at the time.

Size of cells

Cells vary according to whether they are for worker, drone or queen rearing. The worker and drone cells form the familiar "honey comb" **(Plate 2)**. *Worker cells* are smaller than those in which drones are reared and are about 5.25 millimetres across, whereas the larger drone cells are about 6 millimetres across. *Drone cells* are more usually found in the lower parts of the comb, especially towards the corners, but they are frequently built where comb made up of worker cells has been damaged. Worker cells are also used for storing honey and pollen.

Queen cells are built out from the comb. A dozen or more may be made but other colonies may make only one or two. Queen cells are normally only made when the colony is about to produce new queens. A cell is initially formed as a small wax cup and is completed only after an egg has been laid in it. The walls are extended until the cell is shaped somewhat like a peanut shell, hanging down, with an open end. The outer surface is rough. Queen cells are started when the bees want to swarm, supersede (replace the queen), or if the queen is killed.

Space between combs

In the wild, bee colonies construct thick and somewhat irregular combs, suspended more or less parallel to one another and irregularly joined by wax connecting sheets and bars. In man-made hives wooden frames are arranged so that the bees produce vertical combs each consisting of two single layers of cells back to back. It has been found that the space between the frames and the space available for the bees to move through when going around and over the frames or between frames and the wall of the hive is critical. If this space is about 9 millimetres the bees will have enough room to operate efficiently and will tend not to close up spaces with propolis or additional comb and they will be disinclined to build connecting strands or plates of wax from the face of one comb to that opposite it. This results in ease of handling the frames when inspecting the hive or removing the frames for honey extraction. The dimensions of a modern hive and its component parts are based on this critical "bee space". It was the discovery of the significance of the "bee space" which led the Rev. L. L. Langstroth to produce the design on which most modern beehives are based.

Sharing the work

With the great number of different activities which go on inside and beyond the hive, confusion would result if the behaviour of the individual bees was not somehow organised and controlled. The colony achieves this in two ways.

First, certain activities have become the definite responsibility of one or other caste, that is, the labour has been strictly divided among the castes. The queen is responsible for maintaining the population, the drone for fertilising the queen, and the workers have responsibility for all other chores. Secondly, chaos would result if all workers were to feed young or clean the hive or fetch in pollen just when they felt like it.

To ensure that all the tasks are carried out concurrently the colony has evolved a system in which the work done by each individual changes with her age. In this way, with a colony in which there are workers of many different ages, all tasks receive some attention and no vital activity is neglected. The decision as to which task must be done by an individual is not made by her. As she grows older the bee's bodily functions change and her body processes go through a series of changes, so that, for example, when she reaches an age when she should be feeding larvae, her pharyngeal glands develop and she produces royal jelly to feed the larvae; when she reaches an age when she should be making wax, her wax glands develop automatically and her behaviour changes so that she starts to manipulate the wax in the proper way and build comb. There is a natural, controlled sequence of bodily changes and behaviour through which each worker passes during her life from the time she emerges from the pupa until she dies. The best and most efficient sequence has evolved through the ages. The bee is fixed in its way of life and any deviation from it means disadvantage to the colony. If many bees were to deviate the colony would not survive and its behaviour patterns would not be passed on to future colonies, just as inefficient body structures are not passed on because their possessors are less efficient than better equipped bees.

In the description of the castes we saw how each has a body structure adapted to its needs, for example, the queen does not collect pollen and therefore does not have pollen-collecting apparatus on her legs, a drone's eyes are large because he has to pursue the queen during a mating flight, and the worker bee has wax glands to produce the wax from which the comb is built.

The queen's tasks are few and we have already seen that after a few days of wandering in the hive her behaviour changes and she leaves on a few orientation flights. Her behaviour changes again and she sets off on mating flights on which she will be pursued by drones (she is now attractive to them, which she was not during her orientation flights).

Immediately after mating, her behaviour changes again and she returns to the hive. Only a few days after that her ovaries will be fully developed and she will begin going from cell to cell, laying eggs. Another change may occur much later in her life and she may set off with a mass of workers in a swarm to found a new colony elsewhere.

The drones have a simpler life. They are fed by the workers and as their reproductive organs develop they take a few orientation flights. At any time from about three weeks after emergence a drone is able to fertilise a queen and will set off after an attractive queen when she leaves her own hive or when he picks up the scent of a queen in flight from another hive. Successful mating leads to his death. If never successful he will eventually be ejected from the hive by the workers.

The worker goes through many changes of duty during her life because there are a great many tasks to be done in a complex community such as a bee colony. The worker has, broadly, a sequence of three stages in her adult life. At first she is concerned with work related to the brood, she then turns to other activities within the hive, and finally goes out beyond the hive as a field bee. After they emerge from their cells the worker bees' first duty for about two to three days is to clean out the cells from which bees have recently emerged. It is in these clean cells that the queen will lay her eggs. As a good queen is fairly systematic about laying, the young in adjacent and nearby cells tend to be of a similar age and will emerge within a short time of one another. Thus blocks of cells become vacant and ready for use so that the queen does not have to wander about too much in search of cells. A queen which lays irregularly will have her progeny emerging irregularly and this pattern will have to continue so because of the irregular arrangement of available cells next time around. This makes her own work and that of her workers harder. At about three days old, the worker will begin her duties as a nurse bee and she is then responsible for feeding the older larvae. They are fed on a mixture of honey and pollen which the workers take from the cells in which it has been stored. The pollen is partly digested by the bee before being given to the larvae. In general, pollen is stored near the cells containing brood and honey a little farther away, beyond those containing pollen. She tends the old brood for about three days and on about the sixth day, as her pharyngeal glands develop, the worker starts to feed the younger larvae with royal jelly. She makes sure that the small larva, less than three days old, receives an overabundance of royal jelly, so much so that the larva simply lives in a milky looking drop of its food. Older larvae, from the third day, are given less royal jelly but are provided with honey and pollen mixture as required, not in excess. The worker provides according to need. She continues looking after the young larvae until she is about 13 days old. Towards the end of this period she will start taking her first orientation flights. From about the thirteenth to seventeenth day she will be engaged in hive activities and is known as a *house bee*.

Hive activities of the house bee are varied. She will spend some time cleaning out the hive. Removal of old larval skins, bits of wax, rubbish and dead bees is important in reducing risk of disease. A weakened colony which cannot attend to its hygiene runs a greater risk of being attacked by disease and becoming even weaker. Packing away of pollen and nectar into cells for storage is an important duty. A bee bringing in nectar and pollen has little time to waste in the hive—it is more efficient for an experienced pollen or honey gatherer to be out getting on with her job rather than be spending her time looking for storage cells and packing away pollen. So, when she arrives with a load she passes it over to a house bee to take care of and sets off for more. The bees doing house duty are also responsible for "ripening" the nectar, that is, changing it from the nectar as brought in to honey as we know it. This process is mentioned later. During this period the workers engage in comb building. The worker has four pairs of wax glands on the lower side of her abdomen, towards the hind end. When she is producing wax she eats large quantities of nectar as well as pollen and then commences to build comb or cap honey. The wax is produced as a liquid which quickly solidifies owing to complex chemical changes which take place when the wax comes in contact with air. The wax is produced as small scales which the bees manipulate. During the process it is chewed by the bees and at the same time substances from special glands in the head are added which help to make the wax adhere to the comb already in position. The last house activity of the worker is that of guarding the hive. This she does from about the eighteenth to twentieth day. She stands near the entrance of the hive, with her front legs raised a little and her antennae held forwards. The number of guards varies. There are fewer when the colony is very actively bringing in nectar and pollen but when times are harder there are more and any incoming bee will be met and inspected by a guard bee. In an emergency, however, the number of guards is very quickly increased and strange bees have very little chance of gaining access to the hive at such a time. It is known that many bees miss their guard duty period especially when their time happens to coincide with a period of good nectar supply.

When the temperature of the hive rises to about 34°C (the optimum temperature at which the brood develops) the bees will start "fanning". The fanners stand outside or just inside the hive entrance, facing inwards, and vibrate their wings. This causes a strong draught and draws out the warm air. It also removes moisture from the honey, so concentrating it and helping the ripening process.

From time to time young bees take orientation flights before they reach the age at which they become field bees. They usually do this at the warmest part of the day, although orientation flights do take place at other times. The young bees come out of the hive, often large numbers at once, and fly around, gradually becoming more adventurous and going farther from it. During these flights they are learning what the hive and its surroundings look like, in preparation for the time when they will have to find their home with the vital loads of nectar or pollen. These flights are also important hygienically because the bees use them as an opportunity to defecate. Unless they are ill, bees will not defecate in the hive. This habit is a guard against disease which would otherwise soon spread through a dirty hive.

At about 21 days of age the worker is ready to become a field bee. Its life will now be mainly concerned with the outside world and what it can provide for the colony. The colony needs nectar, pollen, water and propolis, and it is the duty of the field bees to find and bring these home. Bees will fly a considerable way to find them and distances of up to 14 kilometres have been recorded, though it is hardly economical for a bee to fly even half that distance. When the resources of nectar and pollen are good most bees will be found within about 750 metres of their hive. In practice, a beekeeper would require all resources to be available within about 1.5 kilometres of the hive if the bees are to be able to provide for themselves and also have a chance to store away additional honey.

Pollen provides the protein requirement in the bees' diet. It provides the body-building material for larvae as well as adults, who need some pollen especially when feeding young larvae, and at wax-making time, although their requirements are less than that of their larvae. The amount of pollen used by a colony varies, of course, according to colony size, brood size, activity and so on. It needs about 10 loads of pollen to provide enough protein to rear one larva and it has been calculated that a colony would need between two to four million pollen loads in a year to supply its requirements. A bee collects pollen deliberately by scrambling over and among the anthers of the flowers so that the pollen adheres to branched hairs on her body and legs. She then wipes it off and collects it in the "pollen baskets" on her back legs. She carries these balls of pollen back to the hive where they are taken over for further processing by the house bees.

Nectar provides energy for the colony, not only the energy used by the bees in moving about but also the energy used in the innumerable chemical processes involved in its bodily functions and in keeping the temperature of the hive at its optimum of 34°C for brood development. Nectar is a sugary solution produced by plants, either in the flowers or from special glands on other parts of the plants, such as at the bases of leaves. The bee is usually attracted to a flower by scent or colour and will suck up any nectar provided into its "honey sac" (crop) in which it is taken back to the hive. Chemical enzymes in the honey sac start to break down the sugars in the nectar. When it arrives at the hive the field bee gives up the nectar, a little to each of several house bees, for storage and ripening. There is much exchange of food and it has been found that in a few hours the nectar from one bee can be spread right through the colony. During its flight a bee will use about one milligram of honey to provide energy to fly one kilometre. For every kilogram of surplus honey a colony will use eight kilograms for its own purposes. The distance flown to obtain this would be equal to about six times round the world!

A good water supply is essential to the colony. It is used to dilute honey when this is being used for food for larvae, and bees probably have to drink water. Water is used to cool the hive in hot weather, with droplets deposited on the combs, closed cells and in other parts of the hive. By evaporation it cools the hive, especially when a draught is created by fanning bees. Water is collected almost every day and one bee may bring as many as 100 loads per day. Bees prefer to collect from shallow water such as a puddle, or water on sand or at the edge of a pond or dam. A colony can use up to 500 millilitres per day, and this is equivalent to about 30,000 loads. The water is given to house bees, just as nectar is, and they retain it until it is needed.

The fourth substance collected by field bees is *propolis*, resinous material produced mainly on the outside of the buds of plants. It is used by the bees for repairing hive damage, sealing off strange objects from the cavity of the hive, partially closing the hive entrance in winter, and sealing down foreign objects which might find their way into the hive, such as sticks, or small animals which have wandered into the hive and died or been killed by the bees. Propolis is carried to the hive in the pollen baskets of the hind legs.

Recognising one another

It is important for members of a colony to know whether a bee which attempts to enter the hive is a member of its own colony or not. It is usually assumed that each colony has an odour of its own which the bees, especially the guard bees, can recognise. This is true, but the matter is not as simple as it may seem. It is known that bees have a small gland on the back of the abdomen which can be exposed and which apparently produces an odour. The bee will sometimes fan her wings when she exposes the gland so that the odour is spread more rapidly. It seems that this is the source of the colony's odour. Experiments have shown, however, that the scent produced is not always the same in the one hive but that it varies from time to time. The main factor in making the change is probably the food source being used by the bees at the time. Unless there is only one food source available, even colonies which are close together will be concentrating mainly on different food sources so that the resulting scents produced in each colony will be somewhat different and will be recognised by all members of the colony as the current scent. It was mentioned that there is considerable food exchange between colony members and that any particular food is soon spread through the colony. This rapid food exchange is probably necessary to ensure that all members of the colony come to have the same scent. Also, if members of a hive are visiting certain flowers for nectar or pollen while those of a neighbouring colony are visiting other species of flowers, they will probably absorb something of the odours of the plants they are visiting and this will probably become a component of the recognisable colony odour for the time being. Strange bees attempting to enter a hive are usually stopped by the guard bees. If the colony is very busy and there is an abundant supply of nectar and honey, there are fewer guard bees, and a stranger can slip in more easily. Also, because there is no "alert" from the guards the house bees take little notice of strangers. If an "alert" is caused, however, such as by banging the hive, the bees of the colony become much more suspicious of strangers and seem to detect them more quickly. It is as though they were suddenly jolted out of their complacent attitude by the banging of the hive. It is interesting that a strange bee which lands in the entrance to the hive by "mistake", that is, without intent to rob, is usually not attacked with violence but is forced away or even carried away by a guard bee. It seems that the behaviour of the intruder is such that the guards recognise that the stranger is not attempting to cause any trouble.

Communicating

If any community is to share out its work load it also has to be able to solve the problem of communication amongst the members. Some of the problems have been solved in the bee colony by making an individual automatically or instinctively do a certain task at a certain time of its life. These periods are controlled by complex chemical changes and processes in the body, involving the development of certain glands and substances which activate them. This, however, would provide a very rigid sequence of events and as the needs of the colony will vary continuously according to the weather, food supply, queen productivity and many other factors, it is essential that the system be flexible so that the colony can change its activities to meet these changing needs. There is, in fact, much flexibility. Although each age group has certain set tasks, it often happens that an individual will carry out duties which are usually done by members of another age group and revert to her more appropriate tasks later on. Some individuals miss out some of the tasks altogether. What they all do really depends on the immediate needs of the colony. Flexibility is essential to colony survival. To achieve this it is essential that the individuals be able to communicate with one another about the state of affairs inside the hive and also in the outside world.

As long ago as 1788 Spitzner, an entomologist in Germany, described how bees carried out a sort of dance when they returned to the hive. The later observations of von Frisch and his colleagues, who carried out research on bees in Austria, provided details of what dances were about. Many researchers have subsequently followed up the work so that we now know in fair detail the nature and function of the famous bee "dances". When a foraging bee returns from the field with a load of nectar which is eagerly taken by the house bees she will probably perform a "dance"; she will go through a series of movements which indicate to other field bees where the source of her nectar is. If the source is within about 100 metres of the hive she will do a "round dance", that is, she will run around on a comb in a circle. When she has completed the circle she will turn around and reverse the direction of her run. Other field bees

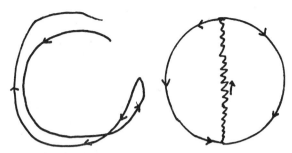

Direction of running in "round" dance (left).
Direction of running in "waggle" dance (right).

will follow her in a generally excited way and in so doing will become aware of odours which the dancer has picked up in the field. They will fly off in random search for a source with that odour. Food is often transferred during this dance from one bee to another so that the other bees will be better able to appreciate the odour or taste of the source. The dancer often dances in several different parts of the hive so that she "recruits" more field bees to work the source she has found. If the source she has found is more than about 100 metres from the hive she will carry out a more elaborate dance and from the way she dances the field bees will be able to tell in which direction and how far from the hive the source is to be found. This dance consists of moving over the comb and circling to right or to left until she has completed a semicircle. She then runs in a straight line wagging her abdomen from side to side until she reaches her starting point. Here she turns in the opposite direction from which she turned on the previous run and again completes a semicircle. This "waggle dance" is repeated over and over. The angle at which she waggles on the comb shows the angle *to the sun* at which the field bee must fly to reach the source. If the dancer waggles her way up the comb the bee must fly directly in the direction of the sun; if the dancer waggles her way down the comb the field bee must fly directly away from the sun. Angles between these indicate the angle to the sun which the field bee must take. In this way the route to be taken from the hive to the source can be communicated very accurately to a new "recruit". This only gives direction, not distance. To indicate the distance of the source the dancer varies the speed of the dance. A fast dance indicates a close source and a slower dance a more distant source. About 10 waggle runs in 15 seconds indicates a source about 100 metres away; only 4 waggle runs in the same time indicates a source about 1000 metres away. So the "recruit" now knows

both direction and distance. Having followed the "instructions" from the dancer a new recruit to that source will be able to find the same source by means of the odour, received from the dancer through food exchange, which the dancer picked up while working at the source. The dances described here are the more obvious ones. There are other dances used to indicate sources less than 100 metres away. It has often been pointed out by apiarists that the dances are carried out in the dark of the hive and that it is not easy to understand how the "recruits" are able to interpret the meaning of the dances. In fact, more recent research has shown that the sounds emitted by the dancers vary and that the dance may be only a result of the actions which the dancer makes when emitting certain sounds; the sounds may well be the important signal to the recruits, not the actual dance.

A field bee which finds a new source of nectar probably measures the distance from it to the hive by the amount of energy used in making the journey. Thus the "distance " indicated by the speed of the dance is really an indication of the amount of energy which the new recruit will have to use to reach the source.

"Communicating" by pheromones

Pheromones are substances produced by one insect to influence the action of another. Queen bees produce a substance called "queen substance". The workers obtain this by licking her during the processes of cleaning and caring for the queen. We know that if enough queen substance is being taken up by the workers and distributed through the colony during food transfer, workers will not build queen cells and laying workers will not develop. If the queen, for some reason such as old age or illness, produces insufficient queen substance or ceases to produce it at all, the effects on the behaviour of the workers is quite dramatic and they start to build queen cells. The workers become aware of the fact that the queen has changed in some way and they take appropriate action by preparing to produce a new queen or queens.

Navigation

Bees navigate in the field by reference to the position of the sun. They are capable of navigation even when it is cloudy and so have some means of determining the sun's position at all times. Field bees are able to compensate for movement of the sun. If they leave the hive and fly in a certain direction the return journey will not be exactly in the opposite direction in relation to the sun because the sun will have moved in their absence. If they are absent for a long time they will have to compensate for the movement to ensure that they end up back at the hive. The sun's movement is not usually so great, however, and once they arrive back in the general vicinity of the hive they can "home in" on the hive using their knowledge of the landmarks they have learnt during their orientation flights and subsequent foraging flights.

Remembering

Bees do have some memory for the sources of food and the position of their hives. They will sometimes go back to a place which has provided a source of nectar for several days after it has ceased to be productive; for example, they will continue to visit a tree which no longer has blossoms producing nectar. In the same way, they will return to a hive position for several days after a hive has been moved.

Swarming

We have seen how the social arrangement of having a reproducing queen, sterile workers and fertile drones ensures the continued existence of the colony. Any species of social insect would, however, soon disappear if it did not have a method of increasing the number of colonies. The honey bee does this by swarming. This is important in the life of the bee and is of great practical importance to the beekeeper. It is discussed in more detail later. Here its significance to the bees themselves is dealt with.

Swarming usually takes place in spring and early summer and consists of the departure from the hive of the queen with a large number of workers. They set off in search of a new site in which to settle. Some time before swarming is to take place drones are produced in numbers. This swarm is called the primary swarm. Prior to this event the workers, in response to reduction of queen substance, start the construction of queen cells in which the queen will lay eggs from which the workers

will rear larvae destined to become new queens. The queen reduces her egg-laying activity and at about the time the new queen cells are closed she leaves the hive with her great retinue of workers. This swarming is accompanied by great excitement and much rushing about in the hive. Some of the workers, known as scout bees, would have been out over the few days before the swarm leaves looking for a suitable site for the swarm to use as a new hive. Before leaving the hive the workers who are to make up the swarm fill their "honey sacs" with honey as food, which will tide them over until they are established in and active from their new hive. This engorging of honey makes them quiet and remarkably docile. Swarms of bees in this condition are very easily handled.

The swarm does not usually go far from the parent hive before settling, usually in a tree. The scouts who have found various possible sites for the new abode dance over the swarm in a similar way to field bees who wish to recruit new workers to a food source. In this way a majority of the bees will eventually accept one of the sites found by the scouts and the swarm will move into the new site and settle down to normal life, the queen laying as soon as suitable comb has been provided. The site chosen is usually a hollow tree or log or some other comfortable cavity, such as a cave, the space in a house wall, above a house ceiling, or a cavity in the ground below a large stone. If no immediate decision is reached on which site is to be used the swarm may start comb construction in the open and in some cases this has been known to become the permanent hive site. Usually, however, such colonies do not survive for more than a season, particularly in cool climates, because the bees cannot easily regulate moisture and temperature conditions in the open.

About four to six days after the departure of the swarm the first new virgin queen to emerge in the old hive destroys any other queen cells and the developing queens inside them. The workers may help her to do this. She mates and becomes the new queen of the old colony. Sometimes, if the old colony is a vigorous one, one of the new virgin queens may also be allowed to leave the hive with a swarm of workers. This swarm is called a secondary swarm, and once it has left the hive behaves in a similar manner to the primary swarm. Departure of a primary swarm, particularly if followed by a secondary swarm (and, perhaps, a less frequent tertiary swarm), obviously leaves the parent colony very much depleted of workers. Although spring is the usual swarming time swarms have been known to leave hives at almost any season other than in cold or bad weather.

Supersedure

When the colony becomes aware that the queen is not functioning normally and well, such as when she is old, her egg-laying capacity is reduced, or she is injured, preparations are made to replace, or supersede, her by rearing another queen. The processes leading to this are probably the result of reduction in queen substance being produced by the inefficient queen. The workers start to build a queen cell and the existing queen lays an egg in the partly formed cell when it is a mere cup, just as in the case of preparation for swarming. Completion of the queen cell takes place after an egg has been laid in it. In a framed hive queen cells for supersedure are usually built on the side of the comb. Few such cells may be built if supersedure is anticipated. The new queen which eventually emerges from the queen cell may allow the original queen to remain active in the hive for some time but eventually she is replaced and the new queen remains as the only functional egg layer in the colony. Through supersedure the maintenance of the population of the colony is assured.

Emergency queen-rearing

We have seen that, under certain circumstances, the workers will prepare queen-rearing cells in which, when the cell is quite small, the queen will lay an egg. The workers then complete the cell and rear a queen which will eventually replace the queen that laid the egg. If, because of some accident, a queen is killed or lost, the workers are still able to replace her provided that there is an egg or a larva younger than three days old available in one of the cells of the hive. If this is so, the workers can enlarge and modify that cell into a queen cell, feed the larva on royal jelly throughout her larval life and so produce a queen. In this case the queen cell will, of course, be somewhere on the comb itself. In supersedure or swarming, queen cells are placed elsewhere, such as near the bottom of a frame. They were built as queen cells from the start and are not modified from other cells. As the destiny of a larva is decided at about three days old, the bees cannot produce queens if larvae younger than that are not available.

The development of the individual bees, their activities and their behaviour, while seeming to be very rigid in their sequence and purpose are, nevertheless, flexible and can be modified by external influences so that they provide for the requirements of the colony from day to day. As the needs and the environment of the colony change the bees can modify their activities to ensure that the best use is made of the resources available; they can also cope with sudden changes which cause emergencies.

Bee food

For food bees need nectar, which they transform into honey, and pollen. From these two sources they can obtain all the food required in their diet to rear healthy larvae, maintain their own bodies, and provide the energy for movement and for carrying out the complicated processes which go on in the body and which we call simply, "life". Like us bees need carbohydrates which they obtain from the sugars in nectar, proteins, minerals, fats and vitamins which they obtain mainly from pollen.

Nectar is essentially a sugary liquid produced by plants. One of its functions is to provide an incentive for insects including bees to visit the plant and in the process to transfer pollen from one flower to another, so ensuring pollination and subsequent seed set. In addition to sugars, nectar contains a little protein, some minerals, a little acid, small amounts of yeast, pollen and bacteria, as well as small amounts of substances which give the honey its characteristic flavours. As these vary from one species of plant to another it is important which plants are used by the bees in determining the final flavour of the honey.

There are many kinds of sugar, simple sugars such as glucose and fructose joining in chemical combination to form more complicated sugars such as sucrose (common sugar or cane sugar). The main sugars in nectar are glucose, fructose and sucrose and they occur in various combinations in the nectar from different plants. Nectar production by the plant is influenced by temperature, humidity, age, and state of the flowers producing it. As nectar is so variable it is inevitable that the resultant honey should vary considerably; this variation in honey is one of the factors which makes it such an interesting food.

Nectar is taken to the hive by the field bees who pass

it over to a house bee for further treatment. They carry it home in the honey sac and a load is anything from 50 to 70 milligrams of nectar. The house bees are responsible for changing the nectar into honey, a process which involves chemical changes in the sugars and removal of water to make the sugar solution more concentrated. The house bee exudes a drop of nectar and water evaporates from it. The process is repeated over and over until a sufficiently concentrated solution results. If water is not removed from the nectar fermentation of the unripened nectar (called "green honey") takes place through the yeasts which are naturally present. The yeast cannot cause fermentation in concentrated nectar or "ripe" honey. In the ripening process nearly two-thirds of the water is removed, i.e., to ripen 50 kilograms of honey about 85 kilograms of water is removed. The process of ripening also involves the breakdown of the sucrose into its component simpler sugars, glucose and fructose. Honey is, therefore, mainly a concentrated solution of glucose and fructose. Honey is stored for future use in honey comb cells which are sealed over by the house bees. When nectar is coming in faster than it can be processed it is stored in the cells in the unripened state to be removed and ripened later when the bees have more time, such as during the night.

Fertilisation in plants is accomplished by the transfer of pollen from parts of the male organs (anthers) to part of the female organs (stigma). The means by which this is achieved are legion and some of the most remarkable mechanisms in nature are those which plants have evolved to ensure pollen transfer. In many cases insects, among them bees, are the agents responsible for the transfer. Pollen is visible as the dust-like material, often yellow or cream, to be found in the flowers and on the bodies of the bees. Some plants produce such copious quantities that clouds of pollen fall from the flowers when they are shaken.

When the bee visits a flower to obtain pollen the grains adhere to the branched hairs on her body. She rakes the pollen together, using her legs, and finally gathers the pollen in the pollen baskets on her hind legs and carries it back to the hive. Here she places the pollen mass in a cell and leaves it for a house bee to take care of. When searching for pollen a field bee usually works

one kind of flower. This is more efficient than searching a variety of flowers because it is more likely that most flowers of a particular species will be releasing their pollen at the same time to ensure cross-pollination of the flowers. Pollens can often be recognised by the pattern of their outer surfaces, which is characteristic for plant species or related groups of species. The house bee mixes a little honey with the pollen so that it becomes a more cohesive mass and stores it in a cell. This moistened pollen and honey mixture is called bee bread and is the food fed to the larvae.

Protein may make up about a third of the dry weight of pollen with about another five per cent being minerals such as potassium, phosphorus, calcium, magnesium and iron. These are all important items in the diet of both larval and adult bees. Although most of the pollen is used for feeding the larvae, adult bees also need it for proper development of the pharyngeal glands (which produce royal jelly) and to enable them to produce wax. Each bee colony needs 30 to 50 kilograms of pollen a year and it requires about 100 milligrams to rear one larva. It takes about 10 to 12 loads of pollen (that is the contents of 20 to 24 pollen baskets) to provide this amount of pollen.

When pollen is short in the field bees will collect other materials which are granular, such as flour or pollard. When bees are seen bringing pollen into a hive it is reasonable to assume that larvae are being reared.

In addition to nectar and pollen, bees will also collect honey dew. Honey dew is the sweet excrement produced by some sap-sucking insects such as scale insects and aphids. These insects take in large quantities of sap and the excrement, which is rich in sugars, falls on the leaves of the plants from which the bees sometimes gather it up. They treat it as they would nectar although the sugars and their relative proportions differ somewhat from those of nectar; honey derived from honey dew has a characteristic flavour. Bees also obtain sugary material from man-made products and in recent years soft-drink containers in garbage cans, especially at picnic sites, have become a source of sugar.

The final "food" requirement of bees is water. They need it to replace body fluids lost during breathing and excreting as well as for the cooling system in the hive.

CHAPTER THREE
Basic Beekeeping Equipment

Success or failure in beekeeping depends on being able to provide a comfortable home for the bees in which they can live in the best possible way, so that their energies can be used efficiently and they can spend their time maintaining the colony properly and storing up an excess of honey, over and above their own needs, which the beekeeper can harvest as his crop. The previous chapter described how bees organise their colony and their way of life. The beekeeper cannot change these. He can, however, make it as easy as possible for the bees to live a normal life and to arrange things so that he can handle and manipulate the colony to check on its health and condition and remove the surplus honey when it becomes available. He does this by housing them in a man-made wooden structure, the "beehive", and by having tools with which to open the hive to examine the contents, remove the honey comb and extract the honey. In addition, he may have other tools to help him introduce new queens, control the queen's movements, collect pollen, feed the bees in times of shortage of nectar or carry out other special kinds of management procedures. This chapter discusses basic equipment; equipment for special tasks will be mentioned as these come up.

Despite the variety of activities which beekeeping might involve it is basically remarkably simple and requires relatively little equipment.

The beehive

The beehive is the most important part of the beekeeper's equipment for it is "home" to the thousands of workers who will spend their time gathering in the honey crop which the beekeeper will harvest. It is best for a beginner to buy a ready-made hive with a colony of bees already in occupation. Not only will he learn to handle and care for his bees if he starts with a "going concern", but he will also see how the hive is made so that he will be able to make up his own additional hives. He will want to make additional hives because there are few beekeepers who are satisfied with one hive once they are used to handling their colonies. Although anyone with a little knowledge and skill in woodworking can make a hive it is almost invariably better if the beekeeper buys his additional hives from a manufacturer. They come in parts, ready cut but neither assembled nor painted. They are not difficult to assemble and instructions are given later for assembly. His first bought, assembled hive will serve as a model and make assembly even easier. The reason for recommending purchase of beehives is that they need to be made fairly accurately because the spaces between the component parts are critical to the comfort of the bees. They have to allow exactly the right "bee space", mentioned in the previous chapter, in which the bees can work. Each hive consists of a bottom board on which stands a box *(hive body)* without top or bottom, in which are suspended rectangular *frames*. This hive body is referred to as the *"brood box"* or *"brood chamber"* as it is the box in which will usually be found most of the brood. On top of this is one or more similar hive bodies, also containing frames, called a *"honey super"* or, more simply, just a *"super"*. The number of these in a hive depends on the state of the colony, time of year and other factors and the beekeeper will alter the number from time to time

A typical beehive.

according to the needs of each colony. Frequently, the hive bodies and frames will be used as parts of different hives, which is another reason for buying ready-made parts. The parts of the hives must all be of standard dimensions so that they are interchangeable between hives. On top of the super (or supers) is a board, the *inner cover*, which is effectively the ceiling of the hive and over this is the *hive cover* or *lid*, which is its roof.

Over the years beehives of many shapes, sizes and designs have been made. Most modern hives are now made on the Langstroth principle, based on the design of the Rev. L. L. Langstroth and incorporating the important dimension of the bee space between the component parts and movable frames. The major variations in design usually involve differences in depth of the hive bodies and the frames they contain as well as in the number of frames in each hive body. Most hives now manufactured will accommodate either 8 or 10 frames. In general I would recommend standardisation on the 8-frame size for the beekeeper with one or a few hives. Handling the hives is sometimes a little awkward and a hive body with 10 frames full of honey can weigh as much as 25 kilograms. For ease of handling the lighter 8-frame hive bodies are probably better. Most hives have what are referred to as "full-depth supers", of which the dimensions are given later. Shallow supers are sometimes used but I would recommend that initially the supers

be full depth, that is, the same depth as the brood box. The following description deals with the components of a standard 8-frame Langstroth hive. The measurements are given in millimetres. Some of the measurements may seem unnecessarily precise; this is because they are conversions from imperial units in which the hives were originally standardised. Manufactured hive components will be tolerably close to the dimensions given.

The bottom board is a board as wide as the hive body (354 millimetres) but preferably 50 millimetres longer than the hive body (508 millimetres plus 50 millimetres). A riser (strip of wood) about 22 by 10 millimetres is nailed to the board along one of the short sides and both long sides. This is the upper side of the board and will be the floor of the hive. On the underside of the board across each short side is nailed a more substantial piece of timber, preferably of hardwood, about 35 millimetres square. These two hardwood blocks form bars on which the hive will rest. The brood box will stand on the risers so that there is a shallow, wide entrance to the brood box and the extra length of the bottom board will make a platform on which the bees can alight and move into the hive entrance. As the bottom board will be near the ground it is more subject to attack by damp, fungus and insects than the rest of the hive. For this reason hardwood is preferred for the supporting bars and the greatest care should be given to preservation and painting of the bottom board.

A hive body, brood box or super is a box made of boards 22 millimetres thick, with outside dimensions 508 by 354 by 241 millimetres. The inside dimensions are, therefore, 464 by 310 by 241 millimetres. The upper edge of the short sides is rebated 11 millimetres to allow metal rabbets to be nailed into the rebate on which the end lugs of the top bars of the frames can be suspended. There are some boxes made with shallow rebates, in which there is insufficient space for metal rabbets, the lugs resting on the wooden edge of the rebate. Rabbets are not popular with many beekeepers because of the extra work in putting them in. It is recommended that the more deeply rebated ends be bought as the metal rabbets make for easier handling when the frames are to be removed and for longer life of the bodies. Well-designed hive bodies are joined by dovetail joints, for strength, and nailed. Nails used for hive construction should be cement-coated or galvanised. Fifty-seven-millimetre, 13 gauge, nails are available for hive-body assembly; at the upper ends where the side is rebated

41-millimetre, 15 gauge, nails should be used. The sizes given here are suitable for brood boxes and supers; shallower supers have the same side dimensions but are not as deep, usually about 135 millimetres, and they carry frames proportionately shallower.

The inner cover is essentially a board which covers the uppermost hive body (usually a super). Its dimensions are the same as the outside length and width of the super so that when it is placed on the super its edges are flush with those of the super. Inner covers can be made of many materials. Some beekeepers use a simple sheet of plywood. This is not recommended as dampness often causes warping and separation of the layers of plywood and does not provide the correct bee space. Inner covers supplied by manufacturers are usually made of some type of fibre board, e.g., masonite, surrounded by a wooden frame, the outer margin of the frame being of the length and width of the super. The frame is usually thicker on one side of the board than the other; the shallower side should be placed downwards when putting the inner cover in position. The space between it and the top bars of the frames in the super will then be correct bee space. Sheets of bituminised material are also adequate as inner covers but are probably less durable than wooden ones and they tend to discolour the wax cappings. Also, as they are not rigid they tend to sink in the middle, reducing the bee space over the centre of the super and hampering the bees' movements. The inner cover often has an opening in the centre in which can be placed a bee escape or through which food can be supplied. This is discussed later.

The hive cover or *lid*, is usually a "migratory" cover or a "telescopic" cover. The main difference is that the migratory cover has outside dimensions the same as the length and width of the super so that it does not extend beyond it, whereas the telescopic cover is wider and longer so that it extends beyond the super and is provided with side and end flanges which overhang a little on the outside of the super. When it is necessary to move hives, as in commercial beekeeping where large numbers of hives have to be stacked for transport to areas where nectar is being produced in abundance, it is more convenient to use the migratory type. Many apiarists with a few hives prefer telescopic covers, as they are less likely to be blown off or otherwise accidentally pushed off.

A telescopic cover consists of a strong board—about 10 millimetres thick is adequate—562 by 408 millimetres. This is attached to a rectangle of wood 38 by 25 millimetres thick, usually dovetailed at the corners, with the same outside dimensions. When placed in position on the super the rectangle should overhang the sides of the super snugly. As the upper surface of the cover is exposed to rain and sun it should be protected on top and around the sides either by bituminised material or, preferably, galvanised metal sheeting, which can be painted. A migratory cover is made in the same way but the dimensions of the board are the same as those of the length and breadth of the super, that is, 508 by 354 millimetres. The wooden rectangle stands on the upper edges of the inner cover. The ends of the rectangle usually have gauze-covered openings drilled through them, of about 20-millimetre diameter, for ventilation. As the telescopic cover overhangs the edge of the super there is no point in providing such

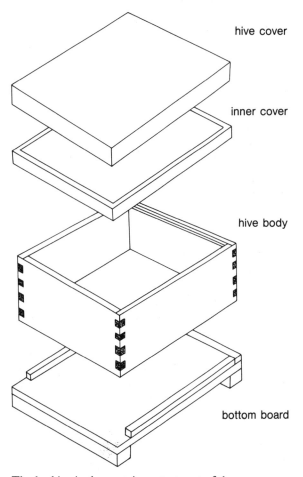

hive cover

inner cover

hive body

bottom board

The beehive is the most important part of the beekeeper's equipment.

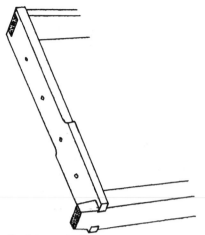

A detail of the end of a frame, showing how the parts are joined.

ventilation ports as they will not open into the air space above the inner cover as they do in the migratory cover.

The frames are the structures in which the bees will build their comb. A frame consists of a top bar and a bottom bar joined by two side bars; the side bars are wider in the upper half than in the lower and the top bar extends beyond the side bars to form two lugs by which the frame hangs on the rabbets at each end of the hive body. The dimensions of the frames, like those of other parts of the hive, are critical. If not of appropriate measurement the bees will tend to stick the frames to the hive body, the inner cover or each other, making it difficult to remove the frames. The reason for having free-hanging frames is that they can be easily removed individually for inspection or manipulation. The top bar is 483 millimetres long by 24 millimetres wide by 17 millimetres deep, except for the end 20-millimetre lug section, which is only 10 millimetres deep. The whole frame is 234 millimetres deep and the bottom bar, which does not extend beyond the side bars, is 445 millimetres. A side bar is 35 millimetres wide in the top half and reduced to 25 millimetres for the lower half. There are four small holes spaced along it. The upper end of the side bar is cut out in the middle to receive the top bar. The parts of the frames are sold separately and the frames are easily assembled by the apiarist. The top bar should be nailed to the side bars using two 31-millimetre, 17 gauge, nails and the bottom bar to the side bar using two 25-millimetre, 18 gauge, nails, all cement-coated or galvanised. Some frames have the top bar lug tapered towards the end but there seems little advantage in this. It will be noticed that the top

end of the side bar is wider than the top bar. This results in the top end of the side bar acting as a spacer which helps to keep the frames the correct distance apart when they are hung side by side along the rabbets. The underside of the top bar may be flat or may have a groove along its length, or a thin pre-cut sliver which can be removed and used as a wedge when putting in the wax foundation, a process which will be described later. Plastic side bars are available into which the top and bottom bars can be inserted. These side bars have a lug at the top so the top bar is not shaped as in the wooden top bars.

Wiring the frames, which consists of stretching a fine wire back and forth across the frame through the holes in the side bars, is necessary before the frame can be used. Before starting wiring, knock a small eyelet into each hole in the side bars; these eyelets prevent the wire from biting into the wood of the frame. The eyelets are knocked in from the outside of the frame. The wires will make the frame more rigid and also provide an attachment for an embossed wax sheet called *foundation*. Special tinned wire should be used.

If only a few frames are to be wired here is an easy method. Hold the frame upright in a vice. Hammer a small panel pin or tack part way into the side bar between the first hole and the top bar. The end of the wire, which comes on a spool, is wound around this a few times, and a length of wire about 2300 millimetres is cut off. Take care not to let the wire on the spool become loose as it is very springy and can easily form knots which can be a nuisance. It is essential to control the unrolling of the piece carefully. The free end of the

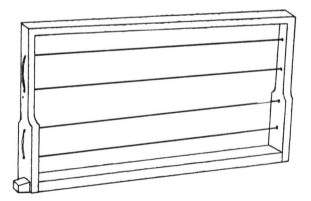

Wiring the frames consists of stretching a fine wire back and forth across the frame through the holes in the side bars.

wire is threaded through the hole nearest to the panel pin and through the equivalent hole in the opposite side bar. The wire is then passed through the adjacent hole in the second side bar and back again to the first side bar. The process is continued until the wire is passed through the final hole, which will be in the same side bar as the first. The wire is made taut by pulling with pliers and finally wound around a second panel pin or tack between the last hole and the bottom bar. The panel pins or tacks are hammered in or bent over to secure the wire. The wire should be tight enough to make a "twanging" sound when plucked.

A piece of apparatus, of which a plan is given below, can be made to wire a large number of frames. To use this, thread the wire loosely from the spool through holes A–H and attach to a tack at position T. Between holes B and C, D and E and F and G run the wire over small wheels (cotton reels are adequate). The frame is held in position on the base board X by a series of blocks M, P, Q, R, S and U which are spaced to take a standard frame. Blocks in different positions will be needed to hold a smaller frame. Having run the wire through the frame holes loosely, tighten it by winding the wire back on to the spool after taking the loop off cotton reel N, then tightening it again after taking it off reel 0, and so on until the wire is taut. Fasten it around another pin or tack at J. A final tightening with pliers may be necessary to make the wire sufficiently taut. It is important to cut off any small ends of wire which protrude from the winding around the pins or tacks as they will be sharp and can be a nuisance when handling the frames later.

Wax foundation sheets are provided as a basis on which the bees can build their comb. Foundation is bought in thin sheets. There are two grades of comb foundation: "thin surplus", which is used where the comb will be removed for eating as comb honey, and "medium brood", which is used for all other purposes including production of honey which will be extracted. The former grade is never wired whereas the latter should be. Foundation is somewhat shorter in length and width than the frame size in use. Each sheet is embossed with a pattern arranged like the arrangement of cells in a honey comb. This sheet is inserted into the frames, supported by the wires, and the bees use it as a "plan" on which to build. The main advantages of the foundation are that it speeds up acceptance of a frame by the bees, saves them the time they would use to start a comb, reduces drone cell formation, and ensures a regular arrangement of cells in the comb. To help you to attach the foundation to the frame you will need an embedding board, which is simply a board 18 millimetres thick and 205 by 428 millimetres, that is, of a size so that the frame can just fit around it. An appropriately smaller board will be needed if a smaller frame size is used.

To attach the sheet of foundation, place it with its long side in the groove in the top bar of the frame and then pass it alternately over and under the stretched wires. If there is a pre-cut sliver of wood in the top bar this is removed, reversed and used as a wedge to hold the long edge of the sheet in place. The wedge is nailed to the underside of the top bar with small nails (12 millimetres is long enough) to secure it and the wax. The sheet is placed alternately above and below the wires as before. It is important to make sure that the upper edge of the wax sheet is in the groove, wedged, or as close as possible to the underside of the top bar depending on top bar design. This will ensure that there is a gap between the other edge of the sheet and the

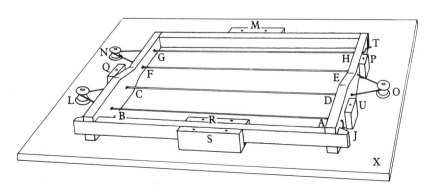

Wiring a large number of frames can be made easier by the use of this simple apparatus.

bottom bar. This will help prevent later bulging of the comb in the lower part of the frame. With the wax foundation in position in the frame the wires can now be embedded in it. This is done by melting the wax along the line of each wire either by using an embedding tool or by an electric current. An *embedding tool* is simply a handle with a double, toothed wheel at the end. The wheel is heated in hot water and run along the wire so that one toothed wheel passes along on each side of it. The heat melts the wax momentarily and the wire is sealed into it as it cools. The wheel must be repeatedly reheated for it to be effective. If using an electric current to embed the wires all that is needed are two wires leading from the terminals of a car battery or a dry cell (6 volts) at the ends of which are attached two nails or other metal contacts which can be used to touch the wires easily. If the wires are touched at intervals, which will be determined by the current and the nature of the contact, they will heat up and almost immediately sink into the wax. This method is very effective, the wires becoming thoroughly embedded in the wax sheet. For the sake of safety low voltages should be used. Mains power can be passed safely through an appropriate 6 or 12 volt transformer to the terminals you are applying to the wires. When embedding, the wires on one side of the sheet are embedded and the frame and foundation are turned over on the embedding board so that those on the other side can be embedded.

It is preferable to place foundation in the frames immediately before use to avoid distortion due to heating between preparation and use.

Using the embedding tool.

A battery or transformer (seen here) can be used when embedding wires in a wax foundation sheet.

Hive timber, preservation and painting. In general, beekeeping equipment costs are not high in comparison with many other hobbies, and as most equipment will be expected to give many years of service a good principle to adopt is that "only the best will do". It will be well worth it in the long run. Even if you eventually decide to give up beekeeping there is usually a reasonable resale value on well-kept equipment. It is especially important that hives and frames be made of good, fault-free timber because they will be out in the open and subject to all temperature and moisture conditions. Also, splits, cracks or knot-holes in the hives cause draughts and uncomfortable conditions, which will result in less efficient working and hence less honey yield. Use only sound hive components. It is worth being a little fussy at the start as broken, warped or misshapen components can later be a nuisance to the beekeeper as well as the bees. If you, as the consumer, insist on good materials, manufacturers will not be inclined to produce poor quality goods. All parts of the hive exposed to the weather should be treated with timber preservative before being painted. It is better to treat the parts to be preserved before assembly so that preservative will penetrate areas subsequently covered, such as the dovetail joints, as these are very prone to trap moisture and decay begins more rapidly at such places. The most often used preservative is a solution of copper naphthanate. Wood treated with this has a greenish colour. Zinc naphthanate is colourless and serves the same purpose. A five per cent solution of pentachlorophenol is an excellent preservative but the wood needs to be brushed down after the preservative has dried before painting At least

48 hours should be allowed between treatment and painting. These preservatives are all painted on or the timber can be dipped in them if suitable containers are available.

When the preservative has dried, nails should be punched well home and the cavity filled with a wood filler. A coat of good-quality primer should be painted on, followed by at least two coats of good-quality outdoor paint. Most hives are painted white; this helps to keep them cool in hot weather. Some apiarists use pastel green or blue and a few have varied the patterns of colour with the intention of helping the bees to locate their own hive. If this is of importance it is probably so only in large apiaries where many hives are standing close together. In any case there is little evidence that bees have lost too much time or effort in finding their own hives in the past. Also, as during the working of hives there is much interchange of hive bodies it may be possible to confuse rather than help bees with this practice. Hives should be painted about every two or three years, especially when there are signs of deterioration in the paint surface. They can be painted while in use but do not paint the landing board at the front of the hive unless you do it at night with a quick-drying paint. Also, make sure that you do not inadvertently use one of the domestic paints, which became available a few years ago, incorporating an insecticide. Along with preparation of new hives and equipment maintenance, painting hives is one of the tasks which can be done in winter when there are fewer pressing jobs awaiting attention.

Stands for hives

If the hive is to stay in one place it is worthwhile making a concrete base on which to stand it. The stand should be a little bigger than the base of the hive and slope slightly towards the front of the hive to assist drainage of rain water away from under the hive. A concrete stand also prevents weed growth near and under the hive, which is an advantage over standing the hive on bricks. Whatever support is devised for the hive it should ensure drainage towards the front and should extend somewhat beyond the limits of the hive itself. Under no circumstances should the bottom board be placed on the ground without cleats as this will shorten its life considerably and make the brood box too damp for the bees' comfort and health.

Smoker, bee brush and hive tool.

Smoker

Apart from the major item of equipment, the hive itself, the essential equipment needed by the beekeeper consists only of a smoker, a hive tool and some protective clothing. The smoker is the traditional tool by which the beekeeper is known. Those not familiar with bees and beekeeping usually visualise a beekeeper as someone wearing a veil and surrounded by a cloud of smoke. In practice, smoke is used sparingly and at very definite times which will be mentioned later when discussing the techniques of handling bees. The smoker consists of a cylindrical metal chamber with a spouted lid and a perforated false bottom from below which a stream of air can be directed into the smoker by means of bellows. Smokers are available in various sizes. Cheaper and light, small smokers do not function for long without needing more fuel and the largest on the market are heavy and sometimes awkward to handle. A medium size is probably the best. As you will be using it in one hand, try out the various sizes by holding a smoker in one hand and pumping the bellows; the biggest size which is comfortable for your hand is the one to use.

Almost anything which will smoulder can be used as fuel and the more smoke it produces the better. Common fuels are pine needles (which have a pleasant smell), old hessian sacking, dry sawdust and bark. To light the smoker, using sacking, take a square of sacking about 250 by 250 millimetres. Hold the middle of the piece with the frayed edges hanging down. Light the edge and when it is smouldering well stuff it into the smoker, close the lid and pump the bellows gently until smoke is produced in quantity. Do not over pump as the sacking will burn with a flame that will produce too hot a smoke, and this will irritate the bees. Pine needles are easily lit by putting a small quantity into the smoker and lighting them so they flame. Before the initial fuel is burnt up add more pine needles, giving an occasional pump of the bellows. When smoke, not flame, is produced in quantity, stuff the smoker fairly full of needles and close the lid. An occasional pump will keep most suitable fuels producing adequate smoke. When bees are affected by smoke they take up honey. This, as in the case of swarming bees, makes them docile and so makes it easy to handle and examine the colony. They are not stupified by the smoke itself so smoking wildly at flying bees is of little avail and only serves to irritate them, although they will avoid smoke. A smoker should always be ready for use whenever bees are to be handled.

Hive tool

The hive tool is one of the beekeeper's most versatile pieces of equipment. Some make do with a chisel or screwdriver but the hive tool is a simple, cheap instrument designed to do specific jobs well. There are slight variations in design but the best is that which is flat and sharpened at one end with a blunt hook at the other and a right-angled notch near the base of the hook. This tool is used for separating hive bodies which have become stuck together, for releasing and lifting frames, and for clearing away propolis and wax. Its uses will

Two types of queen excluder.

be discussed when dealing with the various handling techniques.

Bee brush

When a frame is removed from the hive and the bees clinging to it have to be removed, some beekeepers use a soft brush to brush them off the comb. A certain amount of skill is required to do this without damaging the bees. Also, many beekeepers have experienced a decided tendency to irritation on the part of the bees when the brush is used and prefer to shake the frames or brush the bees away with a deft flick of the fingers. Using a brush means that yet another tool has to be on hand; the smoker and hive tool are essential but the bee brush is not so and many beekeepers do not use one.

Queen excluder

The most usual type of queen excluder is a thin perforated metal plate of the same dimensions as the outside of the hive body. The size of the holes is critical as they are big enough to allow worker bees to pass through but not big enough to allow the queen through. When it is desired to prevent the queen going from one hive body into the next, the queen excluder is simply placed between them. Occasions on which it may be used will be discussed later. Many beekeepers do not have much use for a queen excluder. It is, however, widely used by European, North American and Australian commercial beekeepers. The beginner need not consider it a vital part of his initial equipment and can well wait until he needs it before buying one.

Protective clothing

Many experienced beekeepers do not bother with much in the way of protective clothing when handling their bees. This is because, with experience, it is possible to handle hives gently and without upsetting the colony. The judicious requeening of the colonies with queens from bred stocks which produce docile progeny also makes trouble less likely. Nevertheless, protective clothing should be ready to hand in case it is needed.

A beginner should always wear it, if only to give confidence, until he has learned to handle the bees without likelihood of upsetting them **(Plate 1)**. A pair of white overalls with close (but not too tight) wrist and ankle bands is suitable. Specially designed beekeepers' overalls are obtainable and are probably well worth the money. They have pockets conveniently placed to hold such items as a note book, pencil, matches, bee tool and gloves when not in use. They are made of non-woolly or non-fluff material. Bees do not like fluffy material as their feet are caught up in the fibres and this makes them prone to sting. Tall boots which do not allow easy access to bees creeping up from the ground are useful. A hat is necessary with most designs of veil. The hat should be fairly wide-brimmed to hold the veil away from the face. The veil can be made of almost any material of mesh small enough to prevent the bees getting through but wide enough to be able to see through reasonably easily. Black material is easier to see through than white or green. There are several good

designs of veil on the market and the beekeeper should try them on to see which is most comfortable for him. It is essential that the veil is designed so as not to restrict vision. It should fit well over the hat, or be made so that its upper part is rigid enough to hold the veil away from the face and neck and the lower part is able to be closed, tied around the body or preferably tucked into the top of the overalls, which can then be buttoned up to the neck over it. The beginner should wear gloves which are sting-proof and have the upper part extended long enough to reach well up the forearms where they should be made to fit tightly so as not to allow bees in from the top. With experience gloves are usually the first item of protective clothing to be set aside because even the best make handling frames and tools somewhat clumsy. They should, however, always be within easy reach in case of emergency.

CHAPTER FOUR
Where to Keep Your Hives

The beekeeper with a few hives will probably have them in his own garden or may have a few on a friend's property. The choice of site is often then decided by factors not under his control. The important factors are considered here and the individual beekeeper will have to decide on the site to provide the best possible position. Bees are very tolerant if they have a comfortable hive, so if ideal conditions are not available the beekeeper should not be discouraged—the bees will probably surprise him by producing honey all the same. One of the fascinating things about beekeeping is that almost nothing is certain.

A good nectar source for as much of the year as possible is important. Migratory commercial apiarists move their colonies about the country to take advantage of flowering trees and crops as they become available. Suburban gardens are remarkable in that there always seems to be something to keep the bees going provided the weather is good enough. For surplus honey production the nectar sources need to be within about 1.5 kilometres of the hive.

It is best for the hives to face to the north or north-east so that they receive the morning sun at the entrance. The front of the hive should not face onto a roadway or path or into a neighbour's garden so that the bees fly across it. Sooner or later someone will be stung and the beekeeper will be blamed. Despite the fact that most areas are fully occupied by bees from wild colonies, the nearest known beekeeper is always blamed for the stings. If it is absolutely necessary to face it towards a neighbouring garden, a fence or hedge a few metres in front of the hive will force the bees to fly up and so

probably fly over the danger spot. Otherwise it is preferable to have a clearing in front of the hives so that the bees are not impeded in their coming and going.

A good water supply should be available. If you do not provide one the bees will probably use a neighbour's swimming pool. This will not be popular with the neighbours and in any case you will probably lose a lot of bees through drowning. A shallow container or a tap dripping onto a sand-filled bowl is a better supply. Bees prefer to take water from wet sand, especially if it is in the sun. If a deeper container is supplied it should have a floating block of wood or a sloping wooden or stone ramp from the water so that bees which fall in can find their way out again.

Bees do not like wind so hives should be placed in a protected situation. It is often stated that shade is essential but many apiaries stand in full sun in open paddocks and seem not to suffer any ill effects. As bees work better and are more easily handled when working well it is probably preferable for the beekeeper to keep his hives in a sunny spot if possible. Frames are more easily inspected in bright sunshine when it is easier to see eggs and larvae in their cells. A hive must be placed where it can be worked on easily. It should not be backed by shrubs or a building as it is essential to be able to move freely all around it. Commercial apiaries are usually arranged in rows for ease of access. If there are several hives in one area they should not be closer than about a metre. Bees may "drift", that is, workers may tend to shift so that the strong colonies become stronger as the weaker colonies lose workers to them. If you have a convenient flat roof which will take the weight of you and the hive and is easily accessible, you might consider putting your hive there, if you do not want it in the garden.

Hives should not be too obvious. Unfortunately, there are known cases of hives being stolen and of vandals upsetting hives. If the hives are not too obvious the temptation may be less and neighbours will be less inclined to complain. It is not likely that the amateur beekeeper will have so many hives that he will overstock an area to the extent that the bees find the competition for resources too great. In any case, your bees will be competing with bees from the many wild colonies which occur everywhere.

CHAPTER FIVE
Examining the Hive

How to open a hive

People not used to bees are often amazed when they see an occupied hive opened and completely dismantled. Opening and inspecting a hive to make sure that all is well and to see if there is a surplus of honey which might be harvested is a routine which the beekeeper quickly accepts as normal. If it is done carefully and properly and at the right time it is quite easy. The best way for a beginner to set about getting bees is to buy a ready-made hive, with a colony already at home in it. It will be assumed that this is the case and that the beekeeper now wants to inspect the colony. Other ways of obtaining colonies will be mentioned later. It must be stressed that there is no substitute for experience in opening a hive and every beginner should go to his nearest beekeepers' association to see how it is done and to do it for himself, or get instruction from an experienced beekeeper.

Before opening a hive make sure that you do not smell of any sort of cosmetics; perfumes, after-shave lotions, hair sprays and deodorants often irritate the bees. A wash with plain (not scented) soap and water before starting the operation will remove these and any strong perspiration odour, which bees also find irritating.

Opening the hive should preferably be done in fine, sunny, windless weather, when the bees are most active. A good stream of workers coming and going is a good sign of general activity in the hive. Having put on protective clothing, the smoker should be lit and from time to time during the operation should be checked to make sure that it is ready for use.

Approach the hive from the side or behind, never from in front where you will interrupt the flight of the field bees or interfere with orienting workers. First puff smoke into the hive entrance **(Plate 3)**. A few good puffs from a little way away will suffice if the smoke is dense. If you put the mouth of the smoker right into the opening of the hive the smoke may be too hot. This first smoking will put the guards into confusion and permeate the hive so that the bees fill their honey sacs with honey and become more docile. *Wait for a few minutes* before proceeding. The smoke must have time to pass through the nest and the bees must have time to take up the honey. Working from behind the hive or from the side, take off the hive cover **(Plate 4)** and place it upside down a little way from the hive. It is important to be confident, gentle and slow in all your movements. If you are, there is less likelihood of problems. The bees resent harsh treatment, banging the hive or rapid movements of strange objects around them. The hive tool should be used to loosen parts of the hive which are stuck together. As you take off parts of the hive they should be put out of the way of future operations but not so far that you cannot reach them quickly. A pace or two away is enough.

After taking off the cover the inner cover will still be on the hive, acting effectively as a lid. The inner cover will probably have been sealed to the top super with propolis by the bees. To remove it *gently*, insert the flat end of the hive tool between it and the edge of the super at one corner and loosen the inner cover **(Plate 5)**. As the edge is raised, puff more smoke under it, over the top of the frames which will now be visible. Replace the inner cover for a minute or so before taking it off. Spaces around and between the frames and probably between them and the inner cover will have been filled

1. A beginner should always wear protective clothing.

2. The familiar honeycomb is made up of cells used for the rearing of worker or drone bees and for storing honey and pollen.

3. Smoking the hive entrance.

4. Removing the hive cover.

Burr comb.

with propolis or comb (called burr comb). Burr comb can be gently scraped off with the hive tool and placed in a tin or jar for later treatment. Beeswax is a valuable product and should not be wasted. There will be bees sitting on the underside of the inner cover. They may be left there when the cover is taken off and put to one side or they can be shaken off the inner cover in front of the hive entrance. A sudden downward jerk of the cover will dislodge most of them . They will find their own way back into the hive. If a lot of bees emerge onto the top of the frames to investigate what is going on, a few puffs of smoke across the top of the frames will probably be all that is necessary to send them down again **(Plate 6)**.

The frames can now be removed. Loosen the second frame from one end by inserting the wide end of the hive tool between the first two frames and separating them slightly **(Plate 7)**. Then do the same between the second and third. Repeat the process at the other end of the frame. The second frame should now be free and can be lifted out by placing the curved part of the hive tool under the end lug of the top bar **(Plate 8)**. If it is still stuck the tool can be used as a lever against an adjacent frame so that the second frame is freed. Lift the frame out gently and straight upwards, making sure that bees are not crushed in the process. The frame can now be inspected. The bees can be shaken off in front of the hive or back into it. After inspection the frame can be placed to one side.

It should be noted that the second frame from the wall is removed first. This makes removal of the other frames easier. The end frame can now easily be removed for inspection, followed by the third and fourth. This will leave enough room in the super to move the other frames along after inspection; there is no need to remove them all. After inspecting all the frames they should be replaced *in their original positions*. The whole box can be removed and the frames in the box below inspected

in the same way. The hive tool will probably be needed to release and separate the boxes. A frame must always be held in the vertical plane while being inspected **(Plate 9)**. This is done by inspecting one side first. The comb is then held with one end up, holding the lugs, and turned through 180°. One end is then lowered so that it is upside down but still vertical and with the other side facing the handler. In this way the frame can be examined all over without it being horizontal at any time. If the frame is kept vertical the bees are less disturbed and unripened nectar will not so easily leak out of open cells. Before replacing it the movement is reversed.

The top box should be removed by loosening it at the front of the hive, then at the sides if necessary, and raising the front end **(Plate 10)**. Smoke can again be used over the exposed frames if necessary to quieten the bees. The removed super can be placed on the diagonal on top of the upturned hive cover—this will keep it off the ground and clean **(Plate 11)**. The second super can be placed diagonally on top of this or on the inner cover after inspection.

When inspection is completed the hive must be reassembled in the reverse order so that boxes and frames end up in their original positions, care being taken to space the frames correctly. When replacing one hive body on another it should be placed at an angle to the lower and then slowly turned into position, making sure that bees are not crushed between the boxes **(Plate 12)**. During inspection burr comb found on or between frames should be removed.

What to look for during routine inspection

The beginner can be excused for opening the hives without reason; in any case, mere curiosity is a perfectly adequate and satisfactory reason for opening them. More experienced beekeepers may insist that the hives should be opened only for some better reason because opening causes confusion and upsets the smooth running of the colony; it also changes the temperature and the bees will have to work to re-establish it. If this happens at a busy and productive time it could result in honey loss. On the other hand, the satisfaction of owning a hive or two comes as much from interest and curiosity as it does from the prospect of getting honey, so the hobbyist can afford to indulge himself even if it means a little loss of crop.

From an examination of the colony the beekeeper should have a good idea of the state of affairs at the time and by comparing the situation at subsequent inspections the history of the colony can be followed and any manipulation carried out which might help the bees to better productivity. It is a good idea to make notes on each hive at each inspection; memory is notoriously unreliable and no two colonies behave in quite the same way. This is one of the things which makes beekeeping interesting; there is no fixed routine and each hive has to be managed in accordance with its own history and present state.

Inspection of a normal, healthy hive in summer should show a situation more or less as follows. The brood box should contain eggs, larvae in various stages of development and sealed brood cells. The insides of the cells are best seen by allowing the light to fall directly down them. The brood will be mainly in the lower middle parts of the frames, there being more in the middle frames than in those nearer the ends. Surrounding the brood area will be cells packed with bee bread and pollen in process of being prepared, while surrounding this area will be stored honey. In the supers there should be mainly stored honey, either unripened nectar or capped according to the nectar situation and the use by the colony. In a very active colony the brood may be extended into the lowest super. In the ideal situation the brood, pollen stores and honey stores are arranged in concentric domes in the hive and the frames then appear as though each is a section through the domes. This pattern indicates that the queen is healthy and active. If pollen is being brought in it is highly probable that the queen is active and that there are larvae to feed.

The queen should be sought out and inspected if possible. She is most likely to be found in the brood box, so a start should be made there. If you start the search at the top you concentrate the bees into the lower boxes and the queen becomes even more difficult to find amongst them. Each frame should be systematically searched; do not let your eye be distracted by the general activity on the frame but run your eye across the frames at different levels, then down one side, along the bottom and up the other side. Reverse the frame and repeat the process. Put the frame to one side, preferably in an empty super. Repeat this on each frame and as you remove each frame search the inner side of the brood box well. If you do not find the queen in the brood box replace the frames and put a queen excluder on it. Put

Part of a frame showing capped brood, uncapped, capped drone cells (bottom right) and capped honey (upper part of comb).

the lower super on top of it. A little smoke will help to drive many of the workers down into the brood box. Inspect the now less-populated frames of the super as you did those in the brood chamber. It seldom happens that the queen is higher than the lowest super. The greatest care must be taken during hive inspections to make sure that the queen is not damaged by crushing or lost by dropping from a frame outside the hive. A frame with a queen on it should never be placed outside the hive.

Lack of young brood usually indicates that the queen is lost or not functioning properly and the matter will have to be remedied promptly if the colony is to be maintained. During inspection the health of the brood should be checked. A diseased brood in one hive is a menace to all hives and must be dealt with promptly by reporting it to the Department of Agriculture.

A general inspection is also a good time to clean out a dirty hive. The hive to be cleaned is moved a little to one side and a clean bottom board and brood box placed on its stand. The frames of the dirty hive are cleaned one after the other, starting with those in the brood chamber, and placed in the same relative position in the clean brood box. Greatest care is necessary here not to lose or damage the queen, and the cleaning operation should be combined with a search for the queen. In this way the cleaned frames and their bees are transferred to a clean hive on their own site and their original dirty hive bodies and bottom board can be cleaned in readiness for future use.

CHAPTER SIX
Where to Get Bees

Buying a hive containing a colony

It is best for the beginner to buy a hive containing a colony of bees with which to start operations. The hive may have a single box, the brood box, or may have a super or two as well. In any case, a functioning, active hive will have all the components of a colony and if it is flourishing the beekeeper will be able to examine it and find the various stages of the brood, the stored pollen and honey, the workers and the drones. He can have some practice at finding the queen. He will also be able to watch the comings and goings of the workers as they bring in pollen and nectar and be able to see a complement of guard bees as well as the many bees flying in front of the hive as they become familiar with their surroundings. Colonies are often advertised for sale in local and national newspapers; a perusal of these will give an idea of current prices. Fellow beekeepers at the nearest beekeepers' association may be able to put you in touch with someone with bees for sale and it is a good idea to have someone from the association with you when you go to see your first hive and to help you transport the hive. A complete hive should need only a simple routine handling at first and you can gain experience and confidence in this without worrying about any complicated management procedure. One of the important things about having an active hive is that you can stand by it and watch and get used to the feeling of having bees buzzing around and about you without your becoming worried or frightened. Bees are so fascinating that when you get your first hive you will spend a lot of time just watching them come and go.

Obtaining a nucleus

Another way of starting is to obtain a nucleus. A nucleus is really a small colony. It consists of a few frames with queen, workers, some brood and some comb with pollen and honey to keep the colony going. The beekeepers' association can usually advise where these nuclei can be purchased from bee breeders. The frames are held in a narrow box of the usual hive body depth; the box is made so that it can be sent by rail or otherwise easily transported. When it arrives the nucleus should be put where the colony will finally stand and the entrance opened so that the bees can come out. A little smoke at this stage will quieten them down until they settle in to their new surroundings. After they seem to have settled in, the frames can be moved into a permanent hive on the same spot. As there are only a few frames, probably three, they will not fill the full-sized box, which will be designed to take eight or 10 frames. The frames should be placed together at one side of the hive box and a board placed against the frame nearest the middle of the box. This board functions as a false side to the hive body for the time being. When the bees have settled down an additional frame is put on either side of the original frames which came as the nucleus. The hive is thus built up gradually to fill the whole hive body and the false side is removed. The entrance to the hive should be reduced at first by a slat of wood across the front to help the colony keep warm and make it easier for the bees to defend it. The colony will be small in numbers at first. As the additional frames are put in the entrance can be widened.

Catching a swarm

An easy way to get a colony is to catch a swarm. We have seen that a swarm consists of a queen and some of her workers from an existing colony which are looking for a new home. They are usually docile at this time because they are full of honey. Swarms often settle on low branches of trees or in shrubs within reach of the ground and commonly appear in spring and early summer. If the beekeeper has let it be known amongst friends and neighbours that he would like to catch a swarm he will usually be told if one appears. They are quite spectacular as they fly to a resting place or as they move to their new home. If a swarm is hanging on a branch it is a simple matter to put a cardboard or wooden box under the swarm and shake the branch so that the bees fall into it. Make sure that you take as many of the bees as possible and that the queen is not left behind. Sometimes the branch can be cut so that branch and bees fall into the box. It is essential that the box can be quickly and properly closed so that the bees do not escape on their way to their new home. Bees in a box should not be banged about nor put in the sun. Overheating will cause death. If the box can have a few small holes punched in it (not big enough to let the bees out) so much the better. Even better still is to have part of the wall of the box replaced with fine wire gauze to allow adequate ventilation. Their new hive should have been prepared beforehand and be standing in its permanent place. It should be complete with bottom board, hive body with frames, an inner cover and lid. At least two or three of the frames should contain honey in its comb and, if at all possible, some brood. At least there should be some honey poured over the frames. The brood and honey will encourage the bees to stay when they are "hived", (i.e. introduced to new hive).

If the bees are high in a tree it may be possible to attach a box to a long handle (broom handles joined together end to end by metal tubes or firmly tied to each other will do) and then dislodge the swarm into this.

If you see a swarm and you do not want it for your own use, inform your local beekeepers' association which can contact others who may want it. If you do not someone may well destroy it.

Hiving a swarm

Having caught a swarm in a box it is usually a simple matter to get the bees into their permanent hive. This should be on its permanent site, with a few combs of brood or at least some comb with honey and pollen. A sloping board as wide as the hive is placed in front of the hive with one edge on the landing board of the hive but not so that the hive entrance is obstructed. The swarm is simply shaken out in front of the hive on to the board. In a short time the bees will start to move into the hive. The march is quite spectacular as the bees catch the scent wafted down from those ahead at the hive entrance. As the bees move into the hive keep a careful watch to see if the queen goes in. It may be advisable to place a strip of queen excluder across the entrance for a couple of days (not longer) to discourage the queen from leaving. In most cases the swarm will stay in its new home if brood has been provided.

Removing a wild colony

Removing a wild colony is yet another method of obtaining bees. The colony may be in any situation, such as in a hollow tree, a stump, under a stone or in the cavity walls of a house. The beekeeper will have to use his ingenuity and devise a method of getting the colony into a box. Here are some general principles.

If the colony can be exposed the comb, with brood and food stores, can be cut into sections and tied in position in frames which have not been wired, and placed in a hive. Smoke will be needed before and during the operation. The queen must be taken or at least seen to be present. A match box will hold her until she can be put into the hive. As many workers as possible should be transferred with the cut comb. If the comb is cut in rectangular pieces it can easily be held roughly in place in the frames by rubber bands cut from old tyre tubes or by string or tape. This comb should include brood comb so that the workers will be keen to stay. If the queen and brood are present the bees will probably stay. As many bees and as much comb as possible should be placed in the hive. A slat of wood should be secured across the entrance and the hive taken to its permanent position as soon as possible, where the entrance is opened. When the bees have settled and have prepared comb in all the frames provided, the middle frames can be replaced progressively over a period with good comb and the irregular comb from the wild hive removed, taking away the outer frames first. In time only properly formed frames will remain.

CHAPTER SEVEN

Harvesting Honey from the Hives

As the colonies develop through spring and into summer there will come a time when they gather nectar and convert it into honey at a faster rate than they use it and they will have a surplus to their immediate needs. What the bees are doing, of course, is storing up honey for use in harder times later. This excess is the "crop" of the beekeeper. No matter how many times you take a crop from your hives there is always a feeling of pleasure at the thought of extracting yet another lot of fresh, golden honey. Honey is best taken when there is good honey flow on, that is, when there is an ample supply of nectar in the field. This is not always easy for a beginner to detect but a watch on the build-up of nectar in the frames will indicate whether there is good flow or not. Also, watching the activities of the field bees and the flora at the time will show what is going on.

When the honey crop in the comb is ripe, that is, when it has the desired degree of moisture content and has been changed chemically from nectar to honey, the bees seal over the cells with a shallow dome of wax. This is called "capping". There are several stages in the process of removing honey. In order to take the honey the frames of capped honey are removed from the hive, the capping is cut off and the honey taken out of the cells by spinning the frame in an extractor. The honey is then filtered and bottled and the frames made ready to use again.

Removing the frames

When a super contains a high proportion of capped honey another super can be added to the hive if the honey flow is continuing. This should be placed between the brood box and the full super. Honey will not deteriorate in the capped cells. As long as it is in the hive it will be maintained at correct hive temperature. There is no advantage in removing frames from the hive until they are to be extracted. Honey can be taken at any time but it is more economical in effort if you leave the supers on the hive until there are enough capped frames to make it worthwhile extracting them, remembering that the apparatus has to be prepared before and cleaned after use.

When, by inspecting the hives, it has been decided that there is enough to make a worthwhile extraction, the frames to be extracted are removed. Some frames will probably be fully capped, in others only a proportion of cells will be capped. At least three-quarters of the cells must be capped before a frame is extracted (**Plate 13**). Honey from uncapped cells contains a higher proportion of moisture and when extracted with capped honey it will dilute this to some extent. Such honey is liable to ferment after extraction.

The frames to be extracted have to be cleared of bees. The easiest way to do this is simply to remove the frame from the hive and give it a sudden downward shake in front of the hive, holding the frame vertically (**Plate 14**). With a little practice it is possible to shake off almost all the bees with one sharp, sudden shake. The bees on the ground will find their own way back into the hive. Any which persist in staying on the frame can be gently flicked off with a finger or brushed off using a bee brush. Bees often seem to resent the brush and, in general, it is probably as easy to use a finger. Some beekeepers use only the brush and do not shake the frame at all. Having cleared the frame of all bees it is placed to one side in a box (preferably bee-proof) and the next frame to be taken is dealt with in the same way.

Frames can also be cleared of bees by using an escape board. This is a method which is uneconomical in time and is worthwhile only if you have one or two hives. An escape board is an inner cover or special board of the same dimension with a hole and a device in the middle known as a bee escape. One type consists of a pair of spring wires which converge at one end. They can be bought from beekeeping equipment suppliers. The board is placed between the brood box and the super to be cleared. The bees leave the super through the escape but cannot come back because of the arrangement of the springs. Gradually the super becomes cleared of bees. If the escape board is put in place fairly early in the morning the bees should have all moved out by the following afternoon. The cleared frames can then be removed from the super. It is essential that there be no brood in the supers to be cleared as bees will seldom desert brood.

Supers can also be cleared of bees by using chemicals to drive the bees down. To do this a sheet of metal is cut the same length and width as an inner cover and painted black on one side. To the unpainted side a layer of hessian, felt or other absorbent material is attached. A quantity of the chemical to be used is sprinkled on the lining material and the board is placed, black side up, on top of the super to be cleared. The inner cover and cover are removed beforehand. The black surface will absorb radiant heat from the sun; this will warm the chemical which will evaporate and be diffused down through the super. The bees will move down to get away from it. The chemical most often applied is carbolic acid (phenol), used as a 50 per cent solution of phenol crystals in water. It is not pleasant and must be handled with care. Benzaldehyde and methyl benzoate will also do the job. These chemicals are obtainable from beekeeping equipment suppliers.

It sometimes happens that a few frames remain in a super which do not have enough honey to warrant extraction. These should be left in the super and the full complement of eight or 10 frames in the box made up by putting in frames with foundation or previously used comb. The frames which were in the hive, and which will contain some honey at least, should be moved so that they occupy the middle of the super, and the newly introduced frames should be placed on either side of them. It is more practical to remove frames in

multiples of eight (or 10 according to super size) even if it means leaving a few extractable frames behind.

Frames which have a lot of capped honey and a little closed brood can be extracted, although it will probably result in the loss of the brood. Do not uncap the brood area of the comb before putting the frame into the extractor. Uncapping is explained later.

If the honey flow is continuing when you remove a super for extracting and the colony is still thriving and possibly increasing, a super of frames should replace the super removed; the frames can be some which have previously provided honey and the comb is already "drawn", that is, it consists of old cells, or they can be frames with new foundation. Obviously comb already drawn is to be preferred because the bees can start work on cleaning it out and refilling it with honey straight away without having to rebuild the comb. If a super of frames is not available, replace the combs you have removed for extraction when the honey has been taken. The bees will soon start refilling if nectar is available.

Uncapping

Having taken full frames from the hives they are brought in to be uncapped and extracted. Most hobby beekeepers will do this in the kitchen, the garage or a garden shed, depending on their domestic arrangements (and on tolerance by the rest of the family). The room should be bee-proof if possible because bees will find their way in, attracted by the smell of the honey, and may be a nuisance. Uncapping is the process of cutting off the cappings so that the honey is free to be spun out of the cells **(Plate 15)**. The cappings are removed by using a hot knife. Ordinary knives about the size of a bread knife can be used. The knife is heated by dipping it into very hot water. A pot of hot water on a stove handy to the operator is essential and at least two knives should be available so that when one is being used the other can be heating. The knife must be dried before use.

It is far easier and quicker to use a steam or electrically heated knife, specially designed for the purpose. The electric knife simply plugs into a power point and has an element built into it which keeps the blade hot. The more sophisticated models have this thermostatically controlled.

A steam-heated knife has a hollow blade with an inlet and outlet for steam. The steam comes from a small copper boiler, which is heated on a stove, through a tube to the blade. The steam leaves by another tube which leads to a small tank to receive the condensed steam. The knife should always be kept sharp.

Uncapping should be done over a large basin or wide bucket as there is always a lot of honey attached to the wax cappings. A wooden slat fixed across the top of the basin on which to rest the frame will be useful.

To uncap, the frame is held in the left hand (in the case of a right-handed person) so that one end rests on the wooden slat or edge of the basin. Make sure that you are working at a comfortable height; there is no point in being uncomfortable and making the work harder than it need be. The edge of the knife is run over the cappings, removing them so that they drop into the basin. The more regular the comb, the easier this is. Hold the frame at a slope so that the cut cappings fall away from the comb and not back onto it after cutting. With a little practice long, smooth cuts can be made which result in thin sheets of cappings. Do not cut too deeply into the comb but cut deeply enough to expose the honey. The strokes of the knife can be up or down and sometimes the point of the knife will be needed to uncap irregular parts of the comb. An initial upward stroke of a centimetre or two, at the end of the frame being held, will clear an area from which long downward strokes can be made. When uncapping is completed, the cappings can be suspended in a muslin bag over a basin and the honey, of which there will be an appreciable amount with the cappings, allowed to drain out. The wax cappings should then be set aside for rendering down. Wax is a valuable commodity and throughout beekeeping operations should be kept. Keep even small pieces for later rendering.

Extracting

The honey extractor is the most expensive single item of equipment for the beekeeper. It is used to take out the familiar liquid honey from the comb. This "ordinary" honey is called extracted honey. Extractors come in a wide range of sizes and designs, from the simple "bench extractor" which holds two or a few frames, to large, commercial, motor-driven machines. Most beekeepers' associations have extractors which can be borrowed by members so it is usually not necessary for someone with a few hives to buy his own. By spinning the uncapped frames the extractor flings the honey out of the cells so that the frames can be used

again. This means a great saving in time and energy for the beekeeper and bees, because the bees merely have to clean and repair a used comb before it is ready for use instead of having to start making a new comb.

The bench extractor consists of a large drum, usually of galvanised iron, plastic or stainless steel, in which there are racks which hold frames at a tangent to the sides of the drum **(Plate 16)**. The racks are arranged so that they are revolved by turning a handle on top of the drum. As speed of revolution is increased the honey is flung out of the cells onto the walls of the drum. It runs to the bottom and can be drained off through a fine filter tied over the mouth of the drain or drained into a drum with a fine-mesh filter top.

The frames should be placed in the racks so that the tops of the frames are facing in the direction of spin. This is because cells are not built at right angles to a frame but at a slight angle to it. The speed of revolution is important and has to be learnt by experience. If the frames are not spun fast enough honey is left behind; if they spin too fast the comb will be damaged. Because the honey is slung outwards only one side of the frames can be extracted at a time. If all the honey from one side is removed the weight of the honey on the other side may be too great for the comb to hold and this may be damaged. To overcome this problem about half of the honey can be spun out; the frames are then turned around and the other side spun out completely. The frames are then turned again and the remaining honey spun out from the first side.

The extractor should be firmly held to stop it moving during spinning, especially at the start, before honey has started to accumulate in the bottom. A wooden base with slats nailed to it, between which the base of the extractor fits neatly, can be used. If it is wide enough the operator can stand on the base to give added stability. It is, however, better to have the extractor raised, both for comfort of the operator and so that a container can be placed below the honey outlet.

When it is first extracted honey is slightly cloudy due to particles of wax and small air bubbles. These rise to the surface in time and the wax can then be skimmed off. The honey can be filtered a second time if necessary through a fine filter.

Storage of honey

Even if only a little honey is produced it is unlikely that all will be used immediately. If honey is carefully stored

5. Removing the inner cover.

6. Smoking the tops of the frames.

7. Using a hive tool to loosen frames.

8. Removing a frame.

9. A frame must always be held in the vertical plane while being inspected.

10. Separating hive bodies.

11. Super resting on hive cover during hive inspection.

12. Replacing a hive body.

it will last for a remarkably long time and retain its quality. It must be stored in a container which will not corrode, because although honey is a mild substance it will eventually cause some corrosion. Glass containers with airtight lids are best. A tight-fitting lid is essential to prevent atmospheric moisture getting into the honey. Honey will take up moisture, eventually becoming diluted to the point where fermentation is possible. In time all honey will crystallise. Some honeys crystallise more quickly than others, depending on the source of nectar, the temperature of storage and the age of the honey. This is not to be regarded as deterioration. It is a natural process involving physical and chemical changes in the natural sugars. Crystallised honey can be liquefied by placing the container in hot water. Raising the temperature lowers the viscosity ("thickness") of the honey so that it flows more readily, lowering it increases the viscosity so that the honey becomes "thicker". If honey is kept at low temperatures the tendency to crystallisation increases.

For the small-scale beekeeper it is probably preferable to have a range of sizes of clean, dry glass jars ready at extracting time so that these can be filled straight away. This will be more convenient than having to decant small quantities at intervals from a bulk container. If bulk honey crystallises in a container it is difficult to get it out. The jars can be stored in a cool cupboard. As honey does not deteriorate of its own accord at ordinary temperatures it should not be refrigerated unless particularly viscous honey is required. Also, crystallisation is more likely under refrigeration.

Forms of honey

Extracted honey is the easiest, most useful and most economical form for a beginner to prepare. There are, however, other forms of honey.

Section honey is honey in the comb, sealed and produced in small square or rectangular wooden frames. These are produced in a shallower super, with special dividers between the rows of sections and the sections are provided with thinner foundation so that there is less wax than in normal comb. Section honey can only be produced in a thriving colony during a very strong honey flow when the bees are very short of storage space. There should not be any brood in the sections, so a queen excluder must be used to prevent the queen going into the sections to lay.

Crystallised (granulated) honey is extracted honey which has become granular through the formation of sugar crystals. The more rapid the process of crystallisation has been the finer are the granules. Slowly crystallising honey has a coarse grain. Liquid honey will crystallise if 10 per cent crystallised honey is added to liquid honey.

Cut-comb honey is honey produced in a shallow frame which is not wired. The comb is cut into pieces after it has been capped. In ordinary brood and super comb the foundation is fairly thick because it will be used several times; if the comb is to be cut out a thinner grade of foundation is used. Cut-comb honey is wasteful in wax and effort on the part of both bees and beekeeper but many people find it attractive and prefer to take their honey "straight from the comb".

Chunk honey is extracted honey which has been bottled with a piece of cut-comb honey added to make it more attractive to people who like a little comb with their honey.

Creamed honey is extracted honey which has been beaten to incorporate fine air bubbles into the honey. The bubbles have to be so small that they do not rise in the honey but they enter into a stable physical state with the honey. Ten per cent fine-grain crystallised honey is added to the liquid before beating.

CHAPTER EIGHT

Taking the Wax Crop

The obvious crop from bees is honey and honey production is certainly the main reason most people keep bees. Bees also produce wax and collect pollen. Both are useful substances and should not be overlooked. While the beekeeper with only a few hives would probably not find it worth his while to attempt harvesting pollen he should certainly retain his wax. Sale of wax will contribute somewhat towards defraying his financial outlay on equipment and can also reduce the cost of the comb foundation, which he will need from time to time. It should also be remembered that the bees have used up eight grams of honey to produce a gram of wax. This makes wax quite valuable in terms of the time and labour used by bees to make it.

The main modern uses of beeswax are in candles, (especially those used in religious ceremonies), high quality cosmetics (where it possesses properties which it has not been possible to copy entirely synthetically) and comb foundation for the beekeeping industry. It is also used in dentistry, wood turning, adhesives, inks and chewing gum.

It is advisable for the beekeeper always to have a small box on hand with a tight-fitting lid (a plastic icecream container will do) when he is examining hives. All scraps and bits and pieces of wax should be kept and when an adequate quantity is available, it can be rendered down. The other main sources of wax are from the cappings taken when extracting honey and from old brood and super combs. Cappings wax tends to be contaminated only by honey, which is easily removed, whereas old brood comb contains pupal cases, propolis, cocoon material, dead larvae and other rubbish. Cappings may be contaminated with pollen grains which

are trapped in the wax mainly at the time the wax is used for making the cap. It is better to render the cappings wax separately from that from the combs as the cappings wax is of better quality, but if the brood comb wax is properly and carefully treated it can be made virtually as good.

Wax from cappings will be contaminated by a fair amount of honey, even if methodically drained to obtain the honey. Tied up in a cloth, it should be washed in water to remove the last traces of honey before being rendered.

Brood comb and comb from old supers should be soaked in water to remove honey and to saturate the old cocoons so that they will not absorb wax during the heating process which is to follow. Soaking for about 24 hours is adequate. It is important to use only glass, stainless steel or aluminium containers for rendering. The wax should not come in contact with iron. Enamelled containers can be used but the enamel must not be cracked nor chipped. Contact with iron will darken the wax as does zinc; copper makes it greenish. Wax should be rendered in water, not heated directly in a container, because it is difficult to control the treatment temperature. Wax is inflammable. "Hard" water will sometimes produce a scum-like substance in the wax through reaction between the minerals in the water and some of the substances in the wax. "Soft" water or rain water is preferable. A little vinegar in the water will help to reduce the scum-making reaction.

As beeswax melts at about 64°C, water above this temperature will melt it. A water temperature of 75°–80°C is satisfactory for rendering wax. The simplest way to render small quantities of wax is to place the wax in a porous bag under water. The bag must be weighted down to keep it well below water level because wax is lighter than water. The water is heated; the wax will melt and will rise to the surface of the water. The water must not boil. The bag can be pressed to release more of the wax, while retaining unwanted material. When this is completed the water can be allowed to cool. The wax at the surface will solidify and can be removed as a block. Impure wax on the lower surface of the block can be scraped off. Although simple, a lot of wax is lost by this method because of wax absorption by the old cocoons and the cloth bag.

If an oven is available which has a reasonably sensitive thermostat the bag can be suspended over a dish and the oven turned on. The temperature must not rise above about 80°C. The wax should not be maintained at high temperature for longer than necessary because of chemical changes which might take place and result in lower quality wax. Wax contracts on cooling and rapid cooling will crack the block. It is better to allow the wax to cool slowly so that tensions are not set up in the block.

Rendering equipment using steam or hot water is available from suppliers of beekeeping equipment but the amount of wax produced from a few hives will probably not warrant the expense of such equipment.

A very efficient *solar wax extractor* can be made at very little cost which will render the wax and provide a very clean end product. It consists essentially of a box, with a double-glazed removable lid, in which there is a sloping black tray on which the wax is placed. The extractor is faced towards the sun. The temperature inside the box rises rapidly well above the melting point of the wax, which flows down the tray, through a filter into a removable trough. The wax in the trough is remarkably clean and somewhat bleached by the sun, giving a good quality wax. The rubbish is left behind on the tray or caught by the filter and is removed from time to time. There is little loss using a solar extractor. If the top of the trough is wider and longer than the bottom the block of cooled wax will be more easily removed.

When rendering wax in an oven the receptacle should have sloping sides and a bottom smaller than the rim, so that the wax block can be more easily removed.

If several small blocks of wax have been rendered at different times they can be amalgamated simply by putting them all in one bowl of water and heating to a little above melting point. The wax will solidify on the surface when the water is allowed to cool. The underside of the block will carry some impurities not removed by previous treatments; these can be scraped off.

Wax can be sold directly as a block to manufacturers of wax products or can be sold to makers of wax comb foundation.

CHAPTER NINE

Taking Care of the Bees

So far we have dealt with the hive, its colony of bees and their products assuming that all is absolutely normal, that is, that the colony is docile and has a good queen, there has been a good nectar flow so that the bees have put away a surplus which the beekeeper has been able to take, the bees have remained healthy, and the colony has increased in tune with supplies of pollen and nectar. Inspecting a hive under such circumstances is merely a routine matter to satisfy the beekeeper that all is well. Beekeeping would be rather dull if everything always went along so simply and so well. The interest in beekeeping really comes from the challenges which arise and the decisions which have to be made each time the hive is opened. No two hives are the same. Even colonies started together and standing near each other will soon become different, one may thrive and the other develop slowly. One colony may be docile and easy to handle while another will be much sharper and need more care. Some will fill up the supers in a few weeks, others will go a whole season and not produce enough to warrant an extraction. Encouraging the better hives and trying to help the poorer ones to be more productive are the exciting parts of beekeeping.

To do this the beekeeper will be looking at the weather, checking on the flowers available and on the state of his hives. He also tries to predict what the situation will be like in a week, or a month or in three months' time. Sometimes he is right, often he is wrong; he is seldom sure! This is where the fascination of beekeeping lies. Experience from many generations of beekeepers and scientific research has led to today's beekeeper being able to do certain things at certain times to help his bees. His own skill and experience are all-

important. There is no substitute for experience, his own or that of his fellow beekeepers. Beekeepers are always willing to talk about their bees and exchange experiences. In this way the great pool of knowledge becomes available to everyone. The beginner should not be afraid to ask questions even if he thinks they may be silly ones. Even the most experienced expert was a beginner at one time and no doubt asked the very same questions and felt equally doubtful about asking them.

This chapter deals with some of the things which the beekeeper should be looking for in his hives and what action he should take as a result of what he finds.

Spring inspection

One of the most important times for hive examination is spring as adequate preparation then will have a long-lasting affect on the development of the hive through the rest of the year.

When the weather warms up is the best time for the first spring inspection. There is always the possibility that the weather will turn cold again after the first warm spell. The opening should be delayed until this possibility is over. Knowledge of usual local weather patterns will help make the decision. In general, the colonies will become active with the warming up of the weather. A very active colony, with streams of workers bringing in pollen, is probably in good order with a good queen. At the first spring examination it is important to make sure that there is an active queen. If there is a lot of brood, well arranged in the combs, then all is well. If there are only one or two frames with brood then the queen is probably not functioning well. She may be old or there may be some other reason for her lack of activity. If she is old it may pay the beekeeper to consider uniting the colony with a more vigorous one and building that colony even more rapidly. It would be a waste to unite it with a very vigorous colony which is looking after itself well. The queen may be suffering from lack of space in which case it may be advantageous to provide a super on top of the brood box so that she has more room. It is also essential that there be sufficient stores of honey and pollen for the bees to build up the brood before the main honey flow starts. If they have not had enough stores through the winter they may be short during the vital early spring build-up period. If the onset of warm weather coincides with a dearth of nectar it may be advisable to help the bees over the hard time by feeding them. Methods of doing this are dealt with later. It takes several weeks to produce a field worker bee so it is too late to try to increase your colony when the honey flow starts; your colony will be reaching its best potential when the flow is over.

It is also important to attempt to reduce swarming in spring. By swarming the colony can be reduced to half its size or less and will have a young queen. It will be some weeks before her first offspring are ready to start bringing in nectar, by which time, again, the best part of a honey flow may be over and the bees will have missed their opportunity. Swarm control is discussed later. It is vitally important to have strong, vigorous colonies as early as possible in the season.

It is also important that the bees start the season with a nice clean hive. A dirty bottom board, with piles of old wax, propolis and rubbish which has accumulated over winter is unhealthy and hinders the bees in their activities. The first spring inspection is a good time to make sure that the hives are clean. Cleaning has already been dealt with in an earlier chapter. Finally, the spring inspection can be used to replace any poor or damaged frames. Irregular, old or damaged comb should be replaced with good-quality drawn comb or, if the colony is a very vigorous one, bees can be given frames with new foundation. Combs with an excessive number of drone cells should be replaced. Remember to melt down the wax from the old comb or keep it in mothproof containers until it can be rendered down. It is not advisable to put foundation in a hive which is not very vigorous as the bees will not draw it immediately and they will chew at it to use the wax if they are not producing enough wax for their needs.

Having made sure that the queen is good, that there are stores for immediate use, that there is enough space in which the colony can increase and that there is an adequate field bee population, the beekeeper can allow the bees to get on with their work in preparation for the crop to come.

The next thing to watch for is the time at which to add supers for the bees to fill.

Kinds of supers

As a rule the beginner should use full-depth supers, that is, supers of the same dimensions as those recommended for the brood chambers. A full-depth super full of honey, even with eight frames instead of 10, is quite heavy. Alternatively, it may take a lot of time for the bees to fill such a super if the flow is not intense. In a heavy nectar flow they may fill it quite rapidly. Some beekeepers use shallower supers. These are filled more rapidly than deep supers and are easier to handle but, because of their shallower depth, frames have to be made up which will not be interchangeable with frames from brood boxes. A super with a depth of 180 millimetres is available commercially with frames of appropriate size. We shall assume that the supers to be used are the same depth as the brood box. The principles of "supering", that is, placing supers on the hives, are the same, irrespective of super size.

Queen excluders

Some beekeepers in Australia prefer not to use a queen excluder as it is claimed that the worker bees do not like passing through it and so tend to become crowded below it, even though there is ample room above. By using an excluder the queen is prevented from laying eggs in any box other than the brood box, and the stored honey in the super will not be contaminated by brood. This is clearly an advantage. The usual metal-sheet type of queen excluder may sag in the middle and touch the tops of the frames in the lower box. This results in the space near the middle of the box, above the frames, being reduced. This may deter the bees from behaving normally and make them unwilling to pass through the excluder. There may be less ventilation with a galvanised sheet excluder. An excluder of the rigid wire mesh type is probably better for ventilation.

When to "super"

By watching the activity in the hive, noticing the flowers available to the bees, and by using experience from past seasons, the beekeeper tries to anticipate the need to add a super. To get the best result, it is important that the super be added before it is needed, so that the bees can prepare it, either by cleaning old comb or drawing new comb if the frames have only foundation.

If the bees have started putting wax on the top bars of the frames it usually means that they need room for expansion and they should be given a super.

If the brood box has a lot of brood, sealed or otherwise, and the colony is vigorous, the queen will run out of space for laying and she will reduce her egg output. She must be encouraged to keep on laying and must be provided with the space to do so. A super will do this. It is advisable to try to keep the queen in the brood box. If the bottom box is full and she is given a super she will move up into the super to lay. To encourage her to stay below two or three frames of advanced brood from the brood box can be put in the middle of the newly added upper super and the empty frames from the super put in their place in the middle of the brood box. This region, being the warmest part of the hive, will be used in preference to others by the queen and she will lay in the new frames in the middle of the brood box. The brood in the frames placed in the super will emerge and become part of the workforce and the bees will use the vacated comb for storage. This process can be repeated but the frames of brood as they are moved from below must go into the middle of the super because this is the warmest region and the brood must not be allowed to become cool.

Supers can be added either by "top supering" or by "bottom supering". In top supering the empty super is put on top of the first super. As the supers fill with honey the crop is taken from the lowest super, that is the one above the brood chamber or queen excluder, and the super or supers above it will then be replaced in a lower position. In this system the top supers have to be removed to get to the filled super. In bottom supering the new super is inserted into the hive above the brood chamber or queen excluder and below the filled super. Filled supers are taken off above the more recently added ones. When a super is added in this way it provides space immediately above the brood chamber. The bees have to pass across the newly inserted super to get to the upper super which they were filling before

and so may be more likely to accept the new super quickly. There is always a certain reluctance to invade new supers. Also, by inserting a super the bees are immediately more spread out through the brood chamber and supers as they are interested in what is now the top super because of its honey content. In top supering the bees may be somewhat tardy in starting to use the top super. In top supering, also, there is honey between brood chamber and the newly placed top super which deters the queen from going to the new super to lay. In the case of bottom supering the newly available space has no such barrier and the queen is likely to move into the newly placed super for laying. For this reason it is better to use a queen excluder when bottom supering.

Because of the reluctance of bees to start on new supers it will speed the process up if you place a frame or two with a little unripe honey in the middle positions of the new super. The bees will work out from these. The bees always prefer to work middle frames first. As these are filled and capped they can be moved to the sides of the super and their position taken by frames with less capped honey. As these are now in the middle they will be more rapidly filled and capped.

As long as the weather is warm and the colony vigorous so that it generates a lot of heat, it is advisable to make sure that there is always space available in the hive for brood expansion or honey storage. Only when there is likelihood of cold weather chilling the brood should the bees be allowed to be at all cramped for space.

Fresh air

One of the most important factors in the hive is ventilation. As heat is generated by the activities of the colony, it is essential that the bees can ventilate the hive properly. Rubbish on the bottom board, a small entrance, and comb on top of the frames because of lack of space make it hard for the bees to maintain their hive temperature at its optimum. This is especially so during a honey flow because the bees have to remove considerable quantities of water by forced draught over the nectar as part of the ripening process. They do this by "air conditioning" the hive with a draught of air which can be felt going in one side of the entrance and out of the other after passing over the frames. Anything which interferes with this flow is detrimental to the hive.

Feeding the bees

There are several reasons for feeding bees. We have seen that it is most important that the colonies should be strong and vigorous at the beginning of the honey flow to take advantage of the flow. The growth of the colony should anticipate the flow by some weeks. The onset of a suitable food supply stimulates the queen into laying and ensures that the workers can rear their brood. If food is supplied in advance of the natural honey flow the colony will be stimulated into early growth and development. So, feed the colony in spring to stimulate it, especially if stores are low or exhausted.

If management of the hives has been good, there will be a vigorous colony during the honey flow. The honey flow may decrease between the flowering of one set of plants and another. The queen will respond to this dropping off in flow by reducing her egg output and the colony will be somewhat reduced through the dying off of the older field bees. When the flow starts again the colony will not be at full strength to take full advantage of the renewed flow. During the period between the flows it may be advisable to feed the bees a little even though they are strong colonies. If there is a likelihood that the bees are not accumulating enough stores to see them through winter, they should be fed after the flow is over and before the cold weather sets in. The other times when feeding might be advisable are when you particularly want the bees to be docile or when they may have to be involved in some unusual activity, such as when colonies are being united or when recently collected swarms are being put into their new hives.

The food given to stimulate activity in spring is usually fairly diluted, while that given to a colony between flows or before winter and intended for storage is more concentrated. Bees will tend to use diluted food more rapidly whereas they tend to store up the more concentrated solutions.

What to provide

The usual foods are sugar syrup, dry sugar, candy or honey. The liquid foods are provided in containers of various designs described later.

In all cases only good-quality, refined sugar should be used and honey only from a source which is known to be free of disease.

Dry sugar is placed on top of the inner cover and access given to the bees by making a hole in the centre of the cover. It is essential that a good water supply be available at the same time. Dry sugar is used mainly as a supplementary food for storage and not for stimulating the colony into activity. The sugar can also be placed on papers on top of the frames. In this position it tends to take up moisture from within the hive and become wet, in which condition the bees take it more readily.

Candy for feeding can be made by putting two parts of sugar in one part of water by volume. It should be warmed and agitated until completely dissolved. It is important that there be no free, undissolved sugar. After it has dissolved it must be heated to 135°C. This temperature is fairly critical, as overheating will alter the structure of the sugar and make it unsuitable for candy making. The solution is poured into a flat dish lined with greaseproof paper and allowed to cool. A thin layer about six millimetres thick is best. The slab can be cut up and the pieces put into the hive on top of the frames below the inner cover.

Honey can also be used for feeding but there is a tendency for bees fed on honey to take to robbing other hives. Honey for feeding can be diluted at the rate of one part of water to one part of honey (by volume). If it is provided in one of the feeders designed for use within the hive, such as the Doolittle or Aspden feeder, it seems that the bees regard it as a usual source of honey, not distinguishing it from that available in comb, and are less likely to take to robbing. It is illegal to feed honey in the open under State Apiaries laws.

Sugar syrup is the most usual form of food and is provided at one of two concentrations. Bees tend to store a concentrated solution of sugar, whereas they tend to use for immediate purposes a more diluted solution. If the food is being provided early in the season, to help the colony to start activity in anticipation of a natural honey flow or because stores put by for winter have been prematurely exhausted, then a solution of about one part of sugar to one part of water (by volume) should be used. When the solution is being supplied so that the bees can store it for later use, such as in preparation for winter, a more concentrated solution of about two parts of sugar to one part of water can be used. This stronger concentration will relieve the bees of some of the hard work of water removal during the ripening and storage processes at a time when they will, in any case, be very busy.

Feeders

From what has already been mentioned about the habit of the robbing bees it will be obvious that honey should never be placed out in open containers or in any place where free-flying bees can gain access to it, such as in bird feeders often put out in gardens. Not only is it likely to make the bees inclined to start robbing other hives but there is considerable risk of spreading disease if the honey comes from a diseased hive. It is possible for a honey source to be contaminated by disease and this not be noticed until it is too late and honey already taken to another hive. Honey and sugar syrup should always be provided in a feeder, of which several types are available.

The Boardman feeder is a small wooden box with one side removed and a circular hole made in the top of the box. The hole is wide enough to take the neck of an inverted, wide-mouthed jar. The lid of the jar has a few small holes punched in it. When in use the jar is filled with the food and inverted in the hole. The feeder is then placed at the entrance to the hive with the open side of the box against the nest entrance. By this arrangement the lid of the inverted jar is accessible from inside the hive but not from the outside. Any strange bee wishing to take food from the feeder must enter the hive first, a risky business when the hive entrance is properly guarded.

The Alexander feeder is a wooden or metal trough, as long as the width of the brood chamber and as deep as the distance from the bottom of the brood chamber to the ground, that is, as deep as the bottom board supports. The trough is filled with honey or sugar syrup. The brood box is moved back on the bottom board as far as the width of the feeder so that it projects behind the bottom board. The feeder is placed below the projecting part of the brood chamber, so taking the place of the bottom board at the back of the brood box. If the feeder is made a little wider than the brood chamber and the extra length covered with a removable cover, the feeder can be refilled without disturbing the colony.

The Doolittle feeder is a deep trough, the width of a frame, which hangs in the hive in place of a frame. The feeder is filled with the food. The inside of the vertical walls should be roughened or lined with a rough material so that the bees can gain a foothold and easily climb out of the feeder. This is especially necessary for those bees which accidentally fall into the trough.

The Aspden feeder is a modification of the Doolittle feeder in which, instead of having one deep trough, the frame supports two or three shallow troughs which are divided to form narrow containers for the food. There is less likelihood of bees falling into the Aspden feeder and it is easier for them to get out if they do.

A simple form of Boardman feeder consists of a jar with a push-in or push-on lid. A few small holes are punched in the lid as in the case of the Boardman feeder. Instead of placing the inverted jar at the entrance in a special holder the jar is inverted on top of an inner cover over a hole in the middle. An empty super is then placed on top of the inner cover and the whole closed by the hive lid. The empty space in the super should be filled with crumpled hessian or newspaper or other insulating material. In this way the jar is enclosed entirely within the hive and is accessible only from the top of the frames in the box below the inner cover.

When an inverted container is used, as with the Boardman feeder, the holes in the lid of the jar must be small so that the food does not run out of its own accord but merely forms droplets at the holes which the bees can remove as required. All feeders require fairly frequent inspection and refilling. It is inadvisable to attempt to provide too great a bulk of food at one time because the bees may be slow in taking it and it may have time to ferment in the container. This is more likely when feeding at the more diluted concentrations.

Pollen and pollen substitutes

Feeding honey or sugar syrup early in the season will encourage the bees to take action in anticipation of a honey flow and will encourage the queen to start laying. In practice the beekeeper is attempting to start the colony activities a little prematurely so that the bees' preparations will be well advanced when the natural honey flow comes on. This implies that the hive population will also be increased in expectation of work to come. Brood can, however, only be reared when pollen is available. If this is not readily available in the field it too will have to be provided until it does become naturally available. Fortunately, it is not often necessary to feed pollen under Australian conditions. Pollen can be fed as pure pollen or a mixture of pollen and pollen substitute can be offered.

Pollen can be collected at times when there is an abundant supply and stored for later use. In recent years a market for pollen has been developing, particularly

13. At least three-quarters of the cells on the frame must be capped before a frame is extracted.

14. Shaking bees from a frame at the hive entrance.

15. Uncapping is the process of cutting off the cappings so that the honey is free to be spun out of the cells. This is done by using a hot knife, in this case a steam-heated one.

16. The bench honey extractor flings honey out of the cells so that the frames can be used again.

in some Asian countries where pollen is highly regarded as a food and an encouragement to sexual virility. Pollen is collected in a pollen trap. This is a device available from beekeeping equipment suppliers. It consists essentially of a wire mesh of very specific gauge which forces a proportion of the worker bees to lose their pollen loads as they pass through it on their way into the hive. The pollen falls into a trough from where it is collected, preferably at daily intervals, treated and stored. Pollen can be held in a deep freezer or it can be mixed with fine granulated sugar, one part of pollen to one part of sugar by weight, and stored in tightly sealed bottles. It is important that the pollen and sugar be dry when placed in the bottles and that they are well sealed to prevent moisture getting into the bottle. The pollen can be placed on a fine wire gauze over a lamp until it is dry. An ordinary lamp bulb will generate enough heat. There are times when a colony will store large quantities of pollen in the comb, far in excess of immediate requirements. Such combs can be kept and placed in a hive needing pollen at a later time.

When pure pollen is not available in sufficient quantity or has not been stored, a pollen substitute can be used. In fact, bees often raid unusual and sometimes useless, non-nutritious sources of granular material when they are short of pollen in the field.

A pollen substitute can be made of three parts of soybean flour, one part of yeast and one part of dried skim milk, the parts being measured by weight. If the soybean flour is reduced to about one part, about one and a half parts of casein and half a part of dried egg yolk can be added to make another substitute. The pollen substitute is merely placed near the colony, where it cannot be wet by dew or rain. A flat open container is best and only a little should be offered at a time, that being taken being replaced at frequent intervals rather than a lot being put out at once. If the substitute is not taken soon after it is put out it tends to bind into lumps and is less attractive to the bees.

Perhaps a better way to offer the substitute is to mix it with a sugar solution in the ratio of two parts of substitute (as above), one part of sugar and one part of water. This makes a thick paste which can be placed on top of the inner cover under the hive lid and the bees can get at it through a hole in the inner cover. In this way only the colony for which it is intended will be able to use it and waste by feeding other colonies is avoided.

There is no doubt that bees prefer a substitute which contains real pollen and if possible at least some should be included in the food offered. The higher the proportion of pollen the better it is.

CHAPTER TEN
Swarming, Swarm Control and Absconding

Swarming is the natural way by which bee colonies are increased in the wild. Swarming usually occurs in spring or early summer and in the process the colony giving off the swarm loses its active, functional queen and a large part of its workforce. To the beekeeper this is a decided disadvantage because it is important to have the colonies at maximum strength when the anticipated honey flow starts.

Despite the long experience of generations of beekeepers and a considerable amount of scientific research, it is still not often clear what prompts a colony to swarm when it does. Certainly several reasons are often involved at one time and the beekeeper should be aware of the many causes which could be contributing to swarming in any particular instance. By being aware of these and observing the signs which suggest that swarming might be likely, the beekeeper can often take action to try to prevent swarming.

Causes of swarming

Some strains of bees have a greater tendency to swarm than others and when requeening a colony it is advisable to try to obtain a queen from a strain which is known to be less likely to swarm. Many colonies are, of course, started from swarms collected in the wild. Their origin is often not known but they are not likely to be derived from a "non-swarming" stock.

Production of large numbers of drones seems to encourage swarming or is certainly associated with it. If poorly made or damaged combs are present the bees tend to use these for making drone cells. Anything which encourages drone production should be avoided. It is essential, therefore, that any poor comb in the brood chamber should be replaced by good comb, or foundation which the bees can draw to form good comb.

Presence of queen cells may be a factor in inducing swarming and overcrowding of the hive is probably important. It would seem natural that when conditions in the hive become so crowded that the queen lacks space for laying and the brood cells are full of brood that the colony should make preparations for dividing itself and setting up a new colony by swarming. Anything which makes the hive uncomfortable may also be a factor in inducing swarming, such as continuous overheating of the hive or poor ventilation. Very often poor ventilation may be partly caused by overcrowding. If the brood cells are full of mature brood the workers tend to reduce their activity and stay at home for longer periods. This causes overcrowding and reduces air flow and makes it more difficult for the bees to maintain the correct temperature. The presence of the excess number of bees increases the temperature. The presence of the queen is made known to the workers by the distribution of queen substance from one to the other. This takes time and if there is overcrowding it seems possible that the rate of distribution of queen substance could be slowed down to the point where the workers begin to behave as though there were no queen. This could lead to queen cell production and swarming. In any event, it is important that the beekeeper notes the signs that swarming may be imminent and takes what action he thinks suitable to prevent it.

Indications that swarming is likely

Some time prior to swarming there is usually an increase in the number of drones, after which queen cells are produced. These are usually at or near the bottom of the combs and are sometimes quite numerous. They are built one after the other and the queen lays an egg in each when they are quite small. The cell is then built up by the workers. As a result of this queen cells in a colony preparing to swarm are usually at different stages of development. This is in strong contrast to queen cells produced for supersedure, when the cells are similar in age, are fewer, and are usually built out from the side of the comb. Quite often bees will gather outside the hive entrance, particularly when swarming is associated with overcrowding. This gathering should not be confused with the assembling of bees on the front of the hive in an effort to cool the hive during hot weather, nor with the bees in flight around and over the hive when large numbers of new field bees are orienting. The swarm leaves the parent hive after the queen cells have been capped.

Ways of preventing swarming

Various methods to prevent swarming can be tried and these are directly related to the causes of swarming.

If possible colonies should be provided with a queen which is from a *strain known to be disinclined to swarm*. Colonies originating from a swarm collected in the wild should always be requeened as opportunity arises. The procedure for requeening is quite simple and is described later.

Making sure that the bees have *enough space* for normal activities is important. As bees sometimes are not too keen to pass through a queen excluder it is better not to use one at likely swarming time. Also, if the queen is restricted when she needs additional comb in which to lay, it will not help to have space above the excluder. The comb in the brood chamber must be of worker cells and the combs must be in good condition.

Make sure that the hive is *well ventilated*.

Removal of queen cells may inhibit swarming for a while but usually, once the intention to swarm has reached the stage of queen cell construction, stronger action is probably going to be needed. The bees may well build additional queen cells after removal of the first set and frequent inspection will be necessary, about once a week, to remove later queen cells.

If it is suspected that a colony is about to swarm it can often be prevented by using the *Demaree method*. The frame of brood with the queen on it and the bees adhering to it is removed and placed in another brood box. The box is filled with frames of drawn comb or comb foundation. The box from which the queen was taken is moved to one side and the box with the queen and empty frames put in its place on the original site. A queen excluder is placed on this box and the super (or supers) from the hive placed on top of this. All queen cells must be destroyed. The original bottom brood box (with removed frame replaced by another) is placed on top of the supers. The inner cover and lid are replaced. This manipulation results in a hive with the queen and a little brood and some workers being in the brood box, with lots of space for the queen. Above this are the supers and the remaining brood which was originally in the bottom box, but now separated from the new bottom box by a queen excluder. The queen now has a lot of space in which to lay and the workers will draw out the foundation in the bottom box. The brood in the upper box will be taken care of by the workers as usual and will emerge. That box will become a super. Nine days after the manipulation the brood box should be inspected and any queen cells which might have been started destroyed. After this treatment the bees usually behave as though swarming had taken place and settle back to normal behaviour.

Removal of brood from colonies about to swarm will reduce their inclination to do so. Frames of brood are taken from the hive and distributed to weaker colonies. The frames removed are replaced by frames of drawn comb or foundation. These should be placed at the ends of the box from which the frames have been removed, not in the middle of the brood which is left behind.

Swarm control can also be carried out using the *artificial increase method*. If an increase in colonies is wanted the frames removed as above can be placed, with their bees, in another box and the box filled with frames of drawn comb or foundation. A laying queen can be introduced so that a new small colony is formed. This colony should be fed sugar syrup until it becomes established. It is also advisable to move the small colony some way away from the original strong colony in case the larger colony takes to robbing the new, very weak one. This method combines swarm control with increase in colonies.

Absconding

From time to time an entire colony will leave the hive and settle at a new site. This has some similarity to swarming but does not involve the production of new queens or an increase in the number of colonies. Absconding may be due to uncomfortable conditions in the hive, lack of food or water, or interference by pests such as ants. If the absconding colony is found it can be returned to the hive. It should be fed and any possible source of discomfort, such as insufficient space, poor ventilation of the hive or presence of pests remedied. Long periods of very smoky conditions during bush fires can cause colonies to abscond in an effort to find a smoke-free home even when the colony itself is not threatened.

CHAPTER ELEVEN
Increasing the Number of Colonies

Almost every beekeeper after keeping a hive successfully and after tasting honey from his own hive is keen to increase the number of his colonies. There are several ways of doing this, each needing a different approach.

Most beekeepers take the opportunity to *hive a swarm* when they come across one. This gives another colony. If the swarm has come from his own hive it means that his hive will be considerably weakened by loss of workforce at an important time. It is better to have one strong colony at such a time than two mediocre colonies, hence the importance of trying to prevent swarming. It is not a good idea to rely on swarming to provide more colonies although it is difficult to resist the temptation to hive every swarm you find.

Increase by division can be carried out if the parent colony is strong and vigorous. In this method two or three combs of brood are removed. The combs should contain mostly capped brood with some emerging. It is important to retain the bees sitting on the frames but make sure that the queen is not amongst them. The combs are put into another brood box and a frame or two of pollen and honey put on either side of them. A closed queen cell or a new queen is introduced and the new colony fed with honey syrup. The new hive is removed and placed on a site at least a little way from the parent hive.

An increase in colonies can also be made when *using the Demaree method* of swarm control. When the brood chamber which has been removed from below is placed on top of the super, it is separated from it by a board and raised on three sides with cleats so that it has its own entrance. This should be faced in the opposite direction to the original entrance. A queen cell is allowed to mature in the upper box or any queen cells there can be destroyed and a new queen introduced. This is preferable as it gives an opportunity to introduce a queen from a selected strain. When it is established the new, upper colony is removed and given its own bottom board on a new site.

If *two strong colonies* are available they can be used to provide a third without unduly weakening either of them. Mature, emerging brood frames are taken from the first colony and put in a new brood box. Take the second colony and move it from its place, putting the new brood box on its site. Any bees which are away in the field will come back to the new brood box and supplement the emerging brood. This new colony will not have a queen and a new, selected queen should be introduced to it. The first colony will be reduced in brood but will have retained its queen and field force and be in its original position. The second colony will lose some of its field bees to the new colony, and will have its own queen and brood, but will be in a new position. The new colony will be in the old position of the second hive, and will have brood from the first hive, field bees from the second, and a newly introduced queen. In this way two hives will provide a third but neither will be at a great disadvantage.

If several hives are available the same method can be used but only one brood frame is taken from each. Only one of them is removed to a new site and the new brood box with introduced queen put on its site. In each case, of course, frames removed are replaced by empty brood comb or comb with foundation; these are always placed in the end positions in the box, not in the middle of existing brood frames.

CHAPTER TWELVE
Uniting Colonies

There are times when it is advantageous to unite colonies. This is usually done before a honey flow so that the one strong colony is strong enough to take advantage of it where two or more smaller ones would each be too weak to accumulate excess honey on their own. As it needs about 15,000 bees to maintain the colony, a colony must be in excess of this to accumulate honey. The greater the number above the minimum the better. A colony of 20,000 can field about 5000 field bees, a larger colony of about 50,000 can put about 30,000 into the field.

A colony may be weak for various reasons. The queen may not be vigorous, there may have been disease, the honey flow may have been poor and stronger colonies able to get the major share of the resources available.

Colonies may also be united before winter so that the one colony is more numerous and can keep warmer. When a swarm issues and is recaptured it can be housed for a week or two in a separate hive and then united with the original colony. Whenever uniting is carried out the two colonies involved become confused and are rather defenceless, so it is best for the manipulation to be done late in the day.

When two colonies are to be united it is advisable to feed them both for a week or so beforehand. Inspect both colonies and note the healthier and more progressive. In uniting, one of the queens will have to be destroyed and the better, as judged by the condition of the colonies, should be retained. Because bees do not like allowing bees from other colonies to enter their hive it is important that the uniting process be fairly slow so that both sets of bees become used to the odours of the other and animosity is reduced. If this is not done fighting may break out and many of the bees be killed.

The easiest and most usual method of uniting is quite simple. Remove the queen from the poorer hive. Remove the lid and inner cover from the hive of the stronger and cover it with a sheet of newspaper. A few small holes should be made in the paper. The poorer hive is then removed from its bottom board and placed on top of the newspaper-covered hive. The bees will gradually chew through the paper during which time they will become accustomed to one another and the two workforces will unite without trouble. The frames are rearranged about ten days later so that the brood of both hives are together.

It is possible to unite bees quickly using smoke but the danger of fighting is greater. The better colony is heavily smoked from above and eventually the bees will begin to come out of the entrance in confusion. The bees from the poorer hive, after removal of the queen, are simply dumped in front of the hive amongst the confused bees. They will all then enter the hive together as the smoke dissipates. If they start to fight they should be smoked.

Another method of uniting is to place the better queen, the frames and house bees in a new brood box. Remove the old brood box and put the new one in its place. Then dump the remaining bees from the old brood box in front of the new box and dump the bees from the poorer box, without their queen, amongst them. The combined mass should be heavily smoked or sprayed liberally with sugar syrup. Again, this method is more likely to lead to fighting than the newspaper method.

The inexperienced beekeeper is strongly advised to use the newspaper method. It is slower but safer.

CHAPTER THIRTEEN
Moving Hives

Commercial beekeepers operating on a big scale frequently move their hives to areas in which the honey flow is strong and moving the hives is a highly mechanised process. Instructions here are for moving one or two hives, as almost every beekeeper has to do sooner or later.

Moving a hive can be costly in loss of bees if not carefully planned and carried out. As in all handling procedures with bees, it is important to remember the natural behaviour of bees and treat them without upsetting this. Field bees orientate and make themselves thoroughly familiar with the immediate surroundings of the hive. They return to this small area after each foraging trip and then find their way into the hive entrance. If the hive is moved they will return to the same spot. Different procedures must be adopted according to the distance the hive is to be moved.

Moving a few metres

If a hive is to be moved only a few metres sideways within a garden or orchard it should be shifted a little each day, say not more than about 60 centimetres in a day. The entrance should be kept facing in the same direction until the final new position is reached. In this way the bees returning from the field will be able to find the hive in its new position each day. If the daily move is much longer the bees will become confused on returning to the hive and many may be lost. It is best to make the moves each night when most of the foraging bees are home. They will then set off in the morning from the new position each day and will be less likely

to be confused and waste time when they come back. This waste of time can be important in an active colony.

When the final site is reached the hive can be faced in a new direction if necessary but again the change must be made gradually and the hive rotated only a few degrees each day on its site—a maximum would be about 20 degrees per day. If the hive is being moved backwards or forwards from its old to its new position, the moving distance can be increased to about a metre a day.

Moving up to 1.5 kilometres

Moving distances up to about 1.5 kilometres is a little more difficult. The effective radius of activity from a hive can be about 1.5 kilometres so that the bees from a colony will be familiar with the country within that radius. If moved suddenly to another point in that area they will return to the old site and be lost altogether. The best plan when moving bees within this range is to move them a long distance away, say about 10 kilometres, and leave them there for some weeks to settle down. They are then moved to the desired site. The reason for this is simple. The bees have to reorientate completely in the new site and in the field will not find any landmarks which are familiar to them from the old site and which would guide them back there. After a few weeks their memory of the local landmarks will have been lost and a high proportion of the field bees will never have known the old site at all. The move can then be made to the proper site and they will settle there with minimum loss. This may seem a time-consuming way of moving a relatively short distance but it will save the major part of the field force and prevent a considerable setback to the colony.

Long-distance moves

The bees' memory of places is not a problem where very long moves are made but it is essential that the hives are properly prepared to prevent damage or death of bees. Time and thought spent on this are well worthwhile because many a colony has been lost through inadequate preparation before the move. Loss of bees through physical damage, suffocation, overheating, and incapacity due to the bees being swamped with honey can occur during the move.

Hive boxes should also be specially prepared for the move but, before any noisy or disruptive work is carried out on them, bees should be smoked.

All frames must be secured so that they do not move about or swing inside the boxes. This can be done by pushing the frames firmly to one end of the box and closing up the gaps between them. This will give a gap at one end and the frame at that end can be nailed firmly into position so that it cannot move. Place the nails so that they can be removed easily at the other end of the journey, when the frames are correctly re-spaced. Instead of nailing the end frame, a block of wood can be forced into the gap to hold the frames firmly. If the frames are held by burr comb there is no need to secure them.

As the hive is to be secured so as not to move and will be closed to prevent the bees escaping, it is important to pay attention to ventilation. If this is inadequate the activity of the bees will cause a rise in temperature. The bees keep this down by fanning, an activity which itself generates more heat. In this way conditions soon become lethal to the bees if ventilation is not adequate to reduce the temperature and get rid of the carbon dioxide produced during the activity. The ventilator hole in the middle of the inner cover should be covered by wire gauze and the hive entrance closed by means of wire gauze which can be nailed in position or held there by wooden blocks nailed to the brood box and bottom board. The ventilators at the ends of the migratory lid should be clean and not clogged with propolis, dirt or paint. If the bees are actively gathering nectar they should be allowed to ripen this overnight and be moved in the morning.

It is important to make sure that the parts of the hive are securely attached to one another. Special staples are

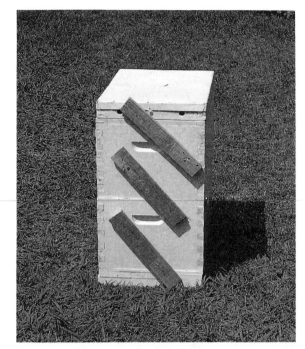

Hive secured by slats ready for moving.

available which can be hammered onto the hive so that one point is driven into the side of the bottom board and the other into the brood box; the same kind of staple can be used to fix brood box to super and super to lid. Usually two of the staples are used on each side to prevent backward, forward and sideways movement. If these staples are not on hand the parts of the hive can be secured to one another by nailing slats of wood from one part to another to prevent movement. A suitable size of slat would be about 10 by 3.5 by 15 centimetres. Each slat should be held in position by four nails and the slats should be placed at an angle to the vertical for strength.

Commercial beekeepers usually strap the parts of the hive together by means of a metal strap which runs over the top of the hive and around under the bottom board. It is put on and tightened by a special piece of equipment. For the beekeeper with only one or two hives, who does not need to move his hives regularly, the expense of the strapping and machine is not warranted and the staple or wooden slat method of securing the hives is quite adequate and efficient.

Bees should preferably be closed in and moved at night because the maximum number of bees will be at home in the hive after dark and it will be cooler during the move. If the journey must include day travel the bees should be kept in the shade when not moving and if the weather is particularly hot the hive should be opened to allow the bees to take a flight and cool the hive; water should be nearby. Hives may even have a little water sprayed into them in a fine spray on exceptionally hot days. In general, however, with only a few hives to move the beekeeper can choose his time for movement so that the minimum inconvenience is caused to the bees who, after all, are the most important participants in the operation.

When the hive is in its new position it should be opened after a little while so that the bees have had a chance to settle down after the movement . If this is during the day the bees which emerge should be smoked to stop them from flying about wildly. This will slow them down and give them time to realise that they are in a new place.

All the operations of packing up and moving should, like all activities with bees, be carried out gently and slowly, with the absolute minimum of banging and jarring of the hives. Bees dislike banging and commotion of any kind. It upsets their routine and whenever they are upset they will not work efficiently. The result of this in the long run is always less honey for the beekeeper.

CHAPTER FOURTEEN

Requeening a Colony

The queen is the most important individual in the colony. It is her duty to maintain the population and she adjusts her egg-laying rate in response to the needs of the colony as communicated to her by the workers, as long as conditions in the hive make it possible for her to do so. Because she is the mother of all the bees in the colony her temperament and qualities will be reflected in her progeny. The drone with which she has mated is also important but we normally have little or no control over which drone will mate with her in the wild. With modern techniques of artificial insemination even this can be achieved, but at a cost which it may not be worth the owner of only a hive or two laying out.

A healthy, properly functioning queen is vital to the production of bees and the gathering of nectar and hence the production of honey. In order to make sure that the colony has a queen which is healthy and able to meet the demands put on her, it is probably worthwhile replacing her with a new queen about every three years although, of course, the natural life of a queen may be much longer than this. In warm climates where the colony does not close down for winter her life will be shorter than in cold climates.

There are a few signs which indicate that a change is due. If eggs are laid on the sides of the cells instead of in the middle it is a sign that the queen is not quite normal. She should have a large, fairly rounded abdomen—if she is flat she is probably deficient. She should be active on the combs in normal circumstances. If she is sluggish she is not adequately healthy or may be getting old. If her laying pattern, as evidenced by the distribution of the eggs and brood over the comb, is a compact semicircular area, she is laying well. If the laying pattern is haphazard she needs replacing.

It is advisable to requeen a colony obtained from a swarm in the field unless you know that the swarm came from a hive with a healthy queen which produced progeny with good temperament and working ability.

Occasionally a queen becomes a drone layer, that is, she lays eggs which all, or mostly, become drones. In this case she should, of course, be replaced.

It is best for the inexperienced beekeeper to buy a queen or queens when needed. The local beekeepers' association will be able to give information on where they can be bought.

Queen bees are sold as "tested" or "untested". In both cases they have been mated but in the latter the breeder has not seen progeny. In tested queens the queen breeder has seen progeny and they can be guaranteed to be true to type. It will usually be adequate for a beginner or someone with only a few hives to buy untested queens, provided they come from a reputable supplier. A beekeeper can also rear his own queens by one of several standard techniques but some experience and a little (inexpensive) equipment is needed. As it is better for a beginner to obtain his queens from elsewhere these methods are not described here.

The queen will be supplied from the vendor in a small cage which will also contain a few workers. There are various designs of cage but in essence they resemble one another in being made of wood and wire gauze and have a quantity of candy enclosed to provide food for the bees on their journey. Such cages are designed for sending through the post without likelihood of harming the bees or their escaping.

When a new queen is to be introduced the old queen should be removed. Unless she is being introduced to a colony which has just become queenless the best time to introduce her is when the season is good. It is not wise to attempt to requeen when there is a chance of swarming or, in general, during the season in which swarming is taking place. There are two basic methods of introducing queens. One is to introduce the new queen gradually so that contact between queen and existing workers is gradual, and the other is to introduce the queen quickly to a confused colony. By far the safer way is to introduce her gradually.

Before introducing the new queen find the old queen and remove or destroy her as well as any queen cells which may be present. It is best to introduce the new queen as soon as possible after the hive has become queenless. It is not advisable to keep the hive queenless for more than about five days and after 12 days it becomes difficult to get the bees to accept a queen. The cage in which she comes will have a hole in one end which is sealed with candy and a cork. The cork is taken out and the cage is placed with the gauze screen downwards on top of the frames over the brood area. There must be contact between hive bees and queen through the gauze. The bees will eat the candy away and in a couple of days the queen will be released. By that time her contact with the bees through the gauze will probably be sufficient to ensure that she is accepted.

Occasionally the bees will "ball" a new queen, that is, the workers will gather around her in a mass and she will eventually be smothered. If this is noticed she should be removed and reintroduced slowly after an inspection to make sure that there are no queen cells in the hive or any other queens present. It is possible that a recently emerged virgin queen might be in the hive. To get the queen free of the "ball" it should be smoked so that the workers disperse.

After requeening the hive should not be opened for about a week; the presence of eggs will indicate that the new queen has been accepted.

CHAPTER FIFTEEN
Getting Ready for Winter

Having had an active season through spring, summer and autumn, the bees will be facing another winter. When discussing the springtime inspection it was pointed out that it is important for the colony to be healthy and to have had ample stores of food through winter. If not, much of its activity will be in re-establishing itself instead of getting on with the work of bringing in the nectar as soon as it is available. A strong colony through winter is essential for a successful season to follow. The time to ensure this is before winter starts. The beekeeper is then looking forward several months to what is to come.

A pre-winter inspection should be routine, no matter how many hives you may have. As they go into winter the colonies must have a strong, vigorous queen because she must be there to start the colony off well in spring. If there are doubts about the quality of the queen she can be replaced in autumn. There must be a large number of workers. It is essential that they be able to maintain the warmth of the hive over winter for their own comfort. If there are not enough bees they will not be able to generate enough heat to achieve this. It may not be worth the trouble of trying to bring a small colony through winter. It may be better to unite a small colony with a larger one. In this way one colony will certainly come through the winter and be stronger at the beginning of activities in spring. This would be better than starting in spring with one mediocre and one small colony as both might be struggling to build up in time for the first honey flow of spring. Although it may seem logical to try to have as many colonies as possible you should always remember that colonies below the minimum size are not going to be very productive. Eight frames well covered with bees are usually needed to bring a colony through successfully in cold areas.

Food is a vital necessity through winter. A good colony needs about 20 kilograms of honey (about eight frames or so) to see it through winter and this should be available. If it is not, supplementary feeding may be necessary. Food is necessary to maintain the hive temperature during winter. It is not wise to take too much honey from the hives in autumn. Pollen should also be available as bees need protein as well as carbohydrates. This is especially important for rearing the last of the summer brood and the first spring brood.

It is important at the pre-winter inspection to make sure that there is no disease in the hive. Having seen that all is in order with the bees themselves it is necessary to ensure that the hive is in good condition and of a suitable size for overwintering. Bees cannot be expected to keep themselves warm in a draughty, faulty hive. Check that all parts of the hive are sound. If necessary reduce the hive to one brood box by concentrating the bees and removing the supers other than the one containing the food supplies. The hive entrance can also be reduced in width to decrease draught. As winter progresses the bees will be using their stores and the greatest danger is in late winter.

It will be obvious that each hive and each area will have to be treated differently. In Australia and New Zealand we have a wide range of climatic conditions and seasonal conditions are very variable.

In some areas the bees will continue to be active throughout the year, although activating warmth does not necessarily mean that pollen and nectar are always available. Sometimes it may be necessary to feed in fairly warm weather to keep the colony going.

In other areas there may be times when there is enough warmth to stimulate the bees into activity for short periods in an otherwise cold winter. This situation is perhaps the most difficult to handle and is the most dangerous. It will lead to more depletion of stores and it may even lead to premature egglaying and an attempt to start brood rearing, with the result that the stores will be even further depleted. If nectar and pollen are available during such short spells all may be well but if not a careful watch will have to be kept on the food supplies. Losses through death of workers will be increased because of their activity and the meagre brood rearing may not make up the loss.

In areas where winters are predictably continuously cold the bees' activity will be reduced for a long period and they will cluster in a mass in the hive. This is an advantage in some ways. The bees will conserve their energies and themselves by not wasting effort on fruitless work. The queen will have a rest from the strenuous activity of egg production and so will probably have a somewhat longer effective life.

With careful pre-winter preparation a populous colony is not likely to be lost and should be ready to take full advantage of what spring has to offer. So often colonies survive winter but are too weak to get off to a quick start in spring because of lack of attention in late autumn. A little attention to preparation for winter will bring a happy result the following season.

CHAPTER SIXTEEN
Honey and Pollen Flora

This book is written for those who have a few hives—it is not a treatise on commercial beekeeping. The two operations are very different. Whereas the commercial beekeeper runs his enterprise as a business, with large capital outlay and an eye to profit, the hobbyist can limit his financial input and replace financial gain with an interesting outdoor activity and a small reward of some "home-grown" honey. A major operation in the commercial beekeeper's annual program is the moving of his hives from one area to another, sometimes hundreds of kilometres apart, to take advantage of nectar flows from plants which he knows, by experience, will be in flower at certain times. The small-scale beekeeper will usually keep his hives at or near his home and is not likely to become involved in frequent long-distance movements.

The sequence of events through the bees' year is dependent on many factors and to take advantage of the nectar flow the beekeeper manages his hives to make sure that they are at maximum strength when the flow comes on. The flow and vital pollen are provided by flowers and the small-scale beekeeper should get to know his local flowers, which are the most generous sources of nectar, and when the nectar flow can be expected. This will help him predict the flow.

Australia and New Zealand are richly endowed with native plants, which produce copious quantities of nectar and pollen. The recent upsurge of interest in native plants for gardens and the increasing practice of local government bodies of using them for landscaping parks, roadsides and other public places has meant that the suburbs are even better places for keeping bees than they were. The aim of beekeeping gardeners should be to plan their plantings to ensure flowers through as much of the year as possible. This is often achieved by using introduced plants to provide flowers at times when the native species are not in flower. Many of the longstanding garden plants are excellent providers of nectar and pollen and help out with supplies, thus spreading the time during which nectar and pollen are available. Weeds are, by their very nature, usually vigorous growers and reproducers. Many weeds also provide nectar and pollen so that even when well-planned gardens fail the bees can still seek out sources which are often quite inconspicuous.

The beekeeper should keep a watch on the local flower scene. It is very useful to make a note of the local species as they come into flower and see which species are used by bees and whether nectar or pollen or both are being collected. Individual bees will use the same source continually until it fades out. Quite often a plant known to produce food will not be visited if another species in the area is already being heavily worked. There may be a change when the first source dries up.

Flowering times vary from place to place and year to year even in the same plant species, so it is the local area, within range of the hives, which is important.

In addition to your own observations you should discuss the local flowering sequence with other beekeepers in the area and with gardeners who will be able to tell you what to expect in an unfamiliar neighbourhood. Some plants are good sources of nectar, others of pollen, some of both.

Generally, eucalypts, as a group, are the most important nectar-yielding plants. This is not to say that they do not sometimes fail or that at times other plants may not be more important. Even in the most heavily built-up city areas there are usually some eucalypts growing in gardens and parks within bee range. The local eucalypt flowering cycles should be carefully watched. As many species of eucalypts are now planted out of their natural areas and many are cultivated for their horticultural value it is worthwhile learning the locally grown species and how to identify them.

In a large genus of plants such as *Eucalyptus*, it is impossible to list all the species which are food sources for bees. The following list includes some of the best and better-known species. It should be noted that the "common names" of plants, especially eucalypts, vary from one area to another. For example, "tallowwood" may refer to any number of different species depending on local naming. One species may have several local names. In the list only one is used. The following are known to be useful to bees: white box *(Eucalyptus albens)*, woollybutt *(E. longifolia)*, spotted gum *(E. maculata)*, flooded gum *(E. grandis)*, yellow box *(E. melliodora)*, ribbon gum *(E. viminalis)*, blue gum *(E. saligna)*, grey ironbark *(E. paniculata)*, lemon scented gum *(E. citriodora)*, scribbly gum *(E. schlerophylla* and related species*)*, grey box *(E. moluccana)*, red box *(E. polyanthemos)*, narrow-leaved ironbark *(E. crebra)*, red ironbark *(E. sideroxylon)*, bloodwood *(E. terminalis)*, tallowwood *(E. microcorys)*, yellow stringybark *(E. muelleriana)*, white mahogany *(E. acmenoides)*, mahogany *(E. botryoides)*, swamp mahogany *(E. robusta)*, peppermint *(E. piperita)*, New England blackbutt *(E. andrewsii)*, narrow-leaved peppermint *(E. radiata)*, blackbutt *(E. pilularis)*, silvertop ash *(E. sieberi)*, yellow gum *(E. leucoxylon)*, red-flowering gum *(E. ficifolia)*, coral gum *(E. torquata)*.

Two species of *Angophora*, *A. costata* (smooth-barked apple) and *A. floribunda* (rough-barked apple) are useful. *A. costata* is erratic in production of nectar but *A. floribunda* regularly produces a large quantity of nectar, which gives a dark, strongly flavoured honey.

The brush box *(Tristania conferta)* is one of the great suppliers of suburban honey because it is so extensively grown as a street tree and garden ornamental.

Despite the profuse flowering of the many commonly planted and beautiful species of bottlebrushes *(Callistemon* spp.*)*, they are not great honey producers, but their flowering periods sometimes occur in a pattern that tides over the bees or stimulates them into activity so that they are ready to take advantage of nectar flows from other species.

Of the paperbarks *(Melaleuca* spp.*)* the best is *M. leucadendron* as it has a long flowering period.

The common tea-tree *(Leptospermum flavescens)* sometimes produces great quantities of nectar and is very attractive to the bees but the honey from its nectar takes a jelly-like form which cannot be extracted properly. The bees can use it for food, however, and it keeps in the hive as well as any other honey, so it can be used by the bees for winter stores.

As a general rule banksias *(Banksia* spp.*)* are not good honey crop producers but they do provide stores for use by the colonies in winter.

Despite the fact that the wattles *(Acacia* spp.*)* are a very common part of the flora and are very conspicuous during flowering time, they are of minor importance in honey production. Their nectar is not produced in the flowers but by glands located elsewhere, usually at the leaf bases or the base of the phyllodes. They mostly produce quantities of pollen and this is their major contribution as far as the beekeeper is concerned.

Because of the enormous number of species of plants which have been introduced and are cultivated or occur as weeds, only some of the obvious species can be mentioned. Local observation will be the best guide to what is important in any particular area.

The pepper tree *(Schinus areira* syn. *molle)* is common, especially in dryer areas, where it thrives, flowers well and is very attractive to bees.

Norfolk Island hibiscus, or white oak *(Lagunaria patersonii)*, is a good provider of nectar and pollen.

Privets *(Ligustrum* spp.*)* produce considerable quantities of flowers which are well worked by bees. Many people find the honey from privets, especially the small-leaved form, unpleasant and too dark.

The willows *(Salix* spp.*)*, especially the weeping willow *(S. babylonica)*, hum loudly with bee activity in spring. Willows are early flowering plants and provide a good stimulus to activity early in the season.

It is well known that the production of commercial crops of fruit is nearly always dependent on the presence of adequate bee populations for pollination. Most fruit trees flower early in the season at a time when bees are building up colonies. Their importance, therefore, lies mainly in stimulation and maintenance of the colonies in preparation for later flows rather than in the accumulation of a honey crop.

Citrus trees can be very useful when fine weather coincides with flowering but some varieties, for example navel oranges, are not very productive.

Although clover is grown mainly as a field crop and bees are important in pollination, some clover plants always occur in non-farming areas, such as on playing fields and roadsides. These areas provide good food supplies for bees but not all species are equally valuable. In some species the flower size makes it difficult for the bees to reach the nectar and they use other more easily reached sources. Red clover *(Trifolium pratense)* is difficult in this way, whereas *T. repens* (white clover) often provides enough nectar for a good crop. It provides excellent quality, light honey.

In country areas the weed Patterson's Curse, or Salvation Jane *(Echium vulgare),* is an important source.

Fennel *(Foeniculum vulgare)* is often found growing in dense stands along roadsides. It provides excellent quantities of nectar which is eagerly collected but the honey has a strong flavour of aniseed which some find disagreeable.

The daisy family (Asteraceae) includes many weed species. As a general rule they produce both pollen and nectar. As their flowering periods, between them, span a large part of the year and they are often common they form a good standby to tide over bees when other sources are reduced.

Camellias *(C. japonica* and *C. sasanqua)* are good winter flowers. The best are the informal varieties which have some stamens producing pollen.

The many varieties of blackberry *(Rubus* spp.*)* are often good producers of nectar, as is lavender *(Lavandula* spp.*).*

With the wealth of flowering species cultivated in the areas likely to be used by the beekeeper with only one or two hives, there is seldom a period when there is nothing for the bees. Even if the nectar flow is not great enough to produce a surplus there are usually at least enough species in flower to maintain the colonies in readiness for the major flow.

CHAPTER SEVENTEEN
Honey and Its Uses

Honey is remarkable for its variety. This is one of its main attractions. No two samples are quite the same. At the same time this variation results in a problem for commercial honey marketers because consumers like to know exactly what the product is like before they buy it and anything different is regarded with some suspicion as being not "right". Not only does the taste of honey vary but all its properties vary. It may be more or less concentrated because the constituent sugars and the multitude of other substances vary in proportions. This all leads to variation in physical and chemical characteristics and the way the honey behaves under different conditions. Although a great deal of research has been carried out and a lot is known about honey, it is not possible to be too precise in giving information about the finer characteristics and reactions of honey in general.

The beekeeper should know at least a little about his honey so that he can guard against spoilage of this hard-earned (by the bees) reward.

Chemical composition

Honey is essentially a super-saturated solution of sugars in water, that is, a solution containing more sugar than the water can usually dissolve. There are various types of sugars. The simple sugars (monosaccharides) can be built up by combining them chemically to form disaccharides (containing two units of monosaccharides) or polysaccharides (more complex sugars made up of many monosaccharides or disaccharides). The higher complex sugars can be broken down chemically into the simple sugars. Carbohydrates are similar to polysaccharides and can be broken down to simple sugars.

Two simple sugars, fructose and glucose, make up about 70 per cent of the honey. Two disaccharides, sucrose (1 per cent) (domestic "sugar") and maltose (7 per cent) make about another 8 per cent and polysaccharides account for another 2 per cent. All the sugars together make up most of the solids dissolved in the water, other substances making up about 2 per cent of the dissolved material. These substances are, however, very important in making honey what it is and giving it many of its most important properties. There are several vitamins, such as B1, B2, B6 and several other members of the vitamin B complex, vitamin A and vitamin C. There are substances which prevent bacterial activity, important in stopping the bacteria from fermenting the sugars. There are several complex substances, enzymes, which are instrumental in causing the breakdown from higher sugars to simpler sugars or vice versa. There are small quantities of alcohols, colouring pigments and substances which give the honey its characteristic flavours, as well as small quantities of minerals. Because of the variety of sources from which the original nectar is collected honey varies in many subtle ways in flavour, colour, consistency and other properties. When bees collect nectar from one source or predominantly one source, the honey may have recognisable characteristics. We can distinguish between yellow box, leatherwood, clover and several other honeys because of the distinctive quality of the original nectar. It is essentially the small quantities of substances, other than the sugar, which provide these characteristics.

Sugars in their pure state always taste the same no matter what the source.

Simple sugars are easily absorbed by the digestive system. In fact, complex sugars are broken down to simple sugars during digestion and the simple sugars taken up. It is because of the high proportion of simple sugars in honey that honey is so easily digested.

Physical characteristics

Honey is heavier than water. A litre of water weighs one kilogram, a litre of honey about 1.4 kilograms. Honey has other properties which are important. It has a high viscosity, that is, it is "thicker" than water and flows much more slowly. It is also "hygroscopic". It takes up or gives out water depending on the amount of moisture in it and the atmosphere. The damper the air the more water is taken up by the honey. By taking up the water the honey becomes more diluted. A point is eventually reached where the honey will not take up any more water. If the honey contains about 15 per cent moisture it will give off and take in water at an equal rate from air which has a relative humidity of about 50 per cent. If there is more water in the honey it will not take water from the air at that higher relative humidity, it will give off water until the water is reduced to 15 per cent. The water is given off or taken in only near the surface so that in dry conditions a dry film will form at the surface which prevents rapid loss of more moisture. Conversely, in a very damp atmosphere the water taken up will remain mainly near the surface so that a fine layer there is more diluted than the bulk of the honey. In extreme conditions fermentation can take place in that film when the honey is so diluted that it no longer stops the fermenting activity of the yeasts that are naturally present in the atmosphere and in the honey. Ripe honey, that is, honey as concentrated and altered by the bees, is normally not fermentable unless diluted.

Crystallisation

Honey crystallises naturally. Crystallisation is not a process of deterioration but merely a change in physical condition. Because it is a supersaturated solution of sugars, the sugars come out of solution easily and form crystals. A small crystal will stimulate this action and

even some impurity, for example, dust particles, will encourage crystallisation. Lower temperatures will encourage it because water is capable of holding less substance in solution at lower temperatures than at higher temperatures. On the other hand, very low temperatures below freezing tend to inhibit crystallisation because they reduce molecular activity in the sugar. The most satisfactory temperature for crystallisation is about 13–15°C. If crystallisation takes place rapidly the crystals are individually smaller and the honey has a finer "grain". Slow crystallisation results in larger crystals and a coarser "grain" to the honey. Crystallised honey is not solid. It consists of crystals of sugars in a liquid medium which is a solution of the other sugars in water. Glucose crystallises much more readily than fructose, and honey which has a high ratio of glucose to water crystallises easily.

Effects of heat

Heat reduces the viscosity of honey, that is, it makes it "thinner" or makes it "run" more easily. Cooling has the opposite effect. Heating will dissolve crystals; these will eventually re-form but crystallised honey can be reliquefied for a long period by heating to 70–80°C for a short while. If the heating is too prolonged or the temperature higher, undesirable changes take place in the honey. The colour will darken irreversibly due to chemical changes in some of the non-sugar components in the honey, especially the small protein content. Some of the sugars will also be altered chemically or broken down. The glucose component is more stable than the fructose component, which is very easily affected by heat. Substances which are aromatic and easily affect our organs of taste or smell, are usually very volatile and volatile substances are easily driven off by heat. As a result, heating honey tends to reduce the aromatic compounds and so alter and generally reduce the taste.

Honey as a commodity

Honey is nature's original sweetener. Humans used honey as their source of sugar long before they learnt to grow or collect sugar from plants such as beet or sugar cane. It is still the main source of sugars for some nomadic peoples and in countries where agriculture and industry are not well developed. Even in industrialised communities where cane sugar has been the main source of sugar for a long time, there has been a dramatic return to honey as a sweetener because of its additional flavours and as part of the general return to natural foods. Honey is ideal for this as processing is minimal between hive and table, no matter in what form it is presented, whether extracted or candied, section or creamed honey. As consumers like to know that they are always getting the "same" product, extracted honey is usually blended. This simply means that honeys of various colours and flavours are mixed to give one of a number of standard colours. Consumers usually have a preference for dark, medium or light coloured honeys. Those who have somewhat more discerning palates and some knowledge of honey may prefer one or other of the "pure line" honeys, that is, a honey which has been predominantly derived from recognisably flavoured nectar from one plant species.

World commercial production of honey is estimated at present to be about half a billion kilograms per year. The main countries producing honey are the Argentine, Australia, Canada, Mexico, Russia and the United States. Much of their produce goes to Europe and Japan. Australia's contribution is about 20 million kilograms, the eighth largest in the world, and there is no doubt that much more could be harvested from the highly productive flora which we have here.

The uses of honey

Honey can be used in almost every way in which sugar is used. It can replace ordinary sugar in drinks and confectionery. It can be used in cooking wherever sugar is included as an ingredient. Its hygroscopic properties are an advantage in cake making as these help to retain moisture and so stop the cake from being too dry.

In many countries honey is considered to have properties in addition to those associated with its sweetness.

It is sometimes used as a medicine and is often a constituent of special invalid drinks. Combined with other ingredients it soothes sore throats. It can be incorporated into soap to make special honey soap and is included in a cream for rubbing onto chapped hands. It is regarded in some countries as having rejuvenating powers and in others as an aphrodisiac.

It is used by athletes to boost the rapidly available energy needed in periods of exertion. Its success in this is due to the high glucose content, glucose being the sugar form used in energy production by the body. Glucose is taken quickly into and through the body.

Its use as a health food replacement for sugar is increasing because in addition to several sugars, it contains small quantities of additional substances not present in refined sugar but considered important, such as the vitamins and proteins.

CHAPTER EIGHTEEN

Diseases, Pests and Ailments

Like all insects, bees have natural enemies in the form of diseases and pests. Some diseases are serious and can be disastrous if allowed to spread. Some pests, too, can be serious but many are only occasionally so or at most are a nuisance. There are a few ailments, mostly due to poor management or harsh conditions, which can cause serious depletion, and even loss of colonies.

Diseases

You should always be on the lookout for disease. Not only will you lose your own stocks but you also have a responsibility to other beekeepers to make sure that your hives are not a source of danger to their colonies. There is legislation to help stop spread of disease but legislation alone is not enough. It requires the full cooperation of all beekeepers, whether they have one or a thousand hives, to make sure that diseases do not spread. Strong healthy colonies are the best protection against disease but even the best-managed colonies may become infected. Any indication that brood or adult bees are dying should be carefully investigated and if in doubt, you should contact the nearest officer of the Department of Agriculture who will be able to put you in touch with the appropriate Apiary Officer. Hives and brood can be infected with more than one brood disease.

Sac brood disease

Sac brood is a disease of the brood and adults caused by a virus. Although young larvae are infected the effects are not apparent until the larvae are fairly advanced. A larva dies after the cells have been capped but before pupating. The cell caps sink in and often have a hole through them. The larva itself becomes discoloured and swollen and full of opaque fluid. The infected larva may lie along the lower side of the cell with the head turned upwards. The dried body can be lifted out (not so in American brood disease—see below). Diseased hives sometimes recover without any action. The virus is airborne or can be taken from a diseased hive to a healthy colony by bees which have been robbing. Special tests are required to confirm the presence of sac brood. Samples of comb or "smears" can be sent to the Apiary Officer for testing. At present there are no actions you can take to eliminate the disease other than making sure your colonies are vigorous.

American brood disease (also known as American foul brood)

American brood disease is caused by a bacterium (*Bacillus larvae*). Infected larvae die and are wet for some time. (If they are touched with a match their remains can be stretched out in a sticky "rope".) Eventually the remains dry out and form a scale-like body stuck to the lower side of the cell. It cannot be removed. Soon after death the larva goes brown, later becoming black. Infected brood will be found scattered amongst apparently healthy brood. Diseased capped brood have the caps sunken and perforated and may have a foul odour, hence the second name of the disease.

Larvae can only become infected by eating the bacteria in the first couple of days after hatching. After the death of the larva the bacteria enter a resting, resistant, spore stage in which they are extremely hardy. They can survive for many years, becoming active and reproductive when they are eaten by a larva. Spores are spread in honey by the house bees and the disease can spread rapidly through a colony. Robber bees will spread the disease from colony to colony. It is easily spread on frames, hive equipment and the hands of the beekeeper. There is no remedial action you can take.

In New South Wales infected hives must, by law, be destroyed by burning. The law in New South Wales also demands that the Department of Agriculture is notified of all cases of American brood disease. Samples of comb can be sent for testing to an Apiary Officer.

Nosema

Nosema is a disease which affects the adult bees, not the brood. It is caused by a minute single-celled animal known as *Nosema apis*. The disease can be detected by a combination of symptoms. The bees have dysentery which can be seen by the appearance of tiny spots on and near the hive. Infected bees have swollen abdomens, cannot fly, and walk around on the ground outside the hive. The wings often hang down and the bee looks generally bedraggled and unhealthy. Numbers of dead bees are found near the entrance to the hive. The colony can diminish very rapidly in bad infections. When in a resistant, resting, spore stage the disease organisms are passed out in the droppings of the infected bees. Other bees become infected by eating contaminated honey or by taking water at a source which is being contaminated by the droppings of infected bees. In this way the disease is easily spread to members of other colonies using the same water source. Beekeeping equipment carrying spores can transmit the disease.

The strictest attention to hygiene at the hive with tools and equipment and at the bees' water source is essential. As usual, general health and vigour of the colony is a necessity if you are to have any hope of saving the colony. Nosema is the most destructive bee disease in Australia.

Positive diagnosis of the disease can only be made by microscopic examination of bee specimens. About 30 live bees can be sent, or preferably taken, to an Apiary Officer for examination. They must be alive so if they are sent by mail they must be provided with candy for the journey.

European brood disease (also known as European foul brood)

European brood disease, like American brood disease, is caused by a bacterium, in this case *Melissococcus pluton*.

Young larvae which are fed infected honey contract the disease and die within a few days. In severe cases the whole colony may die out as a result of the loss of larvae. Large numbers of dead larvae of a few days old indicate an infection of European brood disease. Infected larvae die in unusual postures in the cells and darken, eventually becoming yellowish-brown or dark brown. The larval remains are somewhat similar to those in American brood disease but are not so easily drawn out into a sticky rope, although they are also somewhat foul smelling. All suspected cases of European brood disease should be reported to the nearest Apiary Officer and a sample "smear" sent for confirmation of the diagnosis.

As with most bee diseases an important factor is the general vigour of the colony. Every effort should be made to strengthen the colony, even to the point of uniting colonies to increase the strength. If the queen is of dubious quality and health she should be replaced. As the disease affects the larvae it is possible to reduce the infection by shaking the bees onto a clean, sloping board in front of a hive which is known not to have had infected bees in it and which has frames in it with new foundation. This is not a certain cure as the bees may take infected honey with them. The brood box can be replaced by a new, clean box containing frames with clean foundation and the original brood box placed above it. Eventually the brood in this emerges and the box is then in effect a super with brood rearing going on in the new, lower brood box. After the super is removed for extraction it is fumigated, using glacial acetic acid. The super is placed in an airtight container and about 150 millilitres of glacial acetic acid poured onto an absorbent pad on top of the frames. The container is then sealed and left for at least three days after which it can be aired to remove the acid fumes.

Chalkbrood

Chalkbrood is a fungus disease *(Ascosphaera apis)* which attacks larvae. Dead larvae dry out and are usually white but some may be mottled with blue-grey or black patches. They are not attached to the cell wall (as they are in American foul brood) but are shrunken and have a "chalky" look. The remains of the larvae are often found on the bottom board of the hive and also outside the hive, if the infection is high, because the bees open the capped cells and throw out the dead larvae. The disease is spread by infected honey, pollen and by bees entering strange hives. Infective spores are spread on hive tools and clothing. There is no known treatment at present and the best defence against chalkbrood is strict hygiene and keeping the colonies strong. If there is any suspicion that this disease is present it is essential to contact the nearest Department of Agriculture immediately.

Ailments

Chilled brood

Chilled brood is not a disease but is the result of the brood not being covered adequately enough by the workers to maintain the temperature. This can be caused by lack of vigour in the queen, other diseases which reduce the workforce, or even loss of workers because of insecticide contact in the field. When a colony which has chilled brood returns to strength the workers will clean out the dead larvae. Chilled brood can easily be taken from the cells whereas brood killed by bacterial diseases are hard to remove without breaking them as they adhere to the cell walls. Chilled brood is greyish, whereas in bacterial diseases dead larvae take on a much deeper brownish colour.

Starved brood

Sometimes starvation is the cause of death. There are obviously several causes of starvation, such as lack of pollen or nectar, and reduction in workforce through cold, disease or insecticides. Starved larvae are often taken out of the hive by the workers as opportunity arises. Starved brood are usually white.

Dealing with doubtful causes of death

Doubtful causes of death, whether of adults or brood, should always be referred to an Apiary Officer. If adult bees are involved a sample of a few dozen hive bees should be sent by mail for diagnosis, with a small quantity of candy as food for the journey. If sac brood is suspected a piece of comb, about 12 by 5 centimetres should be sent. The comb should contain brood only, not honey, and should be placed in a cardboard box (not wrapped in plastic, nor in a plastic box) which should be wrapped in brown paper.

If the brood diseases are suspected comb can be sent or a "smear" prepared. A smear is made by placing a few larvae on a glass microscope slide (75 by 25 millimetres) and squashing them. This can be done with a match or small twig. The match is then laid flat on the slide and drawn gently across it so that the squashed larval material is spread out into a thin layer. This is allowed to dry naturally in the air (do not heat) and the slide sent to the Apiary Officer. Small slide-mailing containers are available for this or the slides can be wrapped individually in paper and placed between pieces of thin wood (pieces of plywood cut to a little larger than the slide size are adequate). These can be wrapped in paper and sent to the Apiary Officer. In New South Wales slides and mailing containers can be obtained from local Apiary Officers. Do not allow the match to contaminate another colony. It should be burnt or, preferably, simply left in the hive from which the sample was taken. It is harmless there. A letter of explanation and address of the sender should always accompany or follow a sample for diagnosis.

Pests

There are a number of pests which can attack bees and honey to such an extent that the vigour of the colony can be affected.

Ants

In their search for sweet substances, ants frequently enter hives to steal honey. At the same time they may take brood and they may irritate the adult bees. In severe infestations they have been known to cause the whole colony to abscond. Ants are not easy to control. The hive can be placed on a stand with the feet in a bowl of old sump oil or insecticide. The feet should be placed back from the front of the hive to minimise the risk of returning bees coming in contact with it. Chlordane is the best insecticide available at present for this purpose. The ant nest from which the attackers are coming can be destroyed with insecticide. Always make sure the insecticide does not affect the bees.

Mice

These are sometimes a pest, especially with their attacks on comb in storage during winter when the hives have been reduced in size. They may also attack hives at night and in winter may actually take up residence in a hive. If mouse damage is suspected in winter the hive should be inspected and cleaned on as warm a day as possible.

The entrance to the hive should be reduced by wire mesh which will allow the bees to come and go but which is too narrow for the mice.

Introduced European wasp *(Vespula germanica)*
This is a potentially serious pest of bees. Although no instances of actual damage have yet been reported in Australia its depredations in some other countries, such as New Zealand, are serious. The wasp lives in colonies, often in a cavity in the ground, and is about the size of a bee but with a strongly contrasting pattern of black and yellow. The wasps enter hives, especially late in the season, to remove honey. They are capable of eliminating bee colonies by killing bees during their raids. These European wasps are a potential new pest to Australian beekeepers. Specimens were first seen in 1959. They have been in New Zealand for a much longer period where they cause considerable losses. In the last few years increasing numbers of nests have been found in the Sydney area. When nests are found they should be reported to the Department of Agriculture. Specimens suspected of being European wasps can be sent to the Australian Museum for identification.

From time to time *other insect pests* may cause damage. Occasionally dragonflies will take bees and mantids often prey on bees as they visit flowers. Robber flies take bees in flight as do scorpion-flies, but these pests are usually not serious as the number of bees they are capable of taking is limited.

Among the most important potential bee killers are the *insecticides* used for plant protection. Bees may come in contact with these by directly flying through the spray or by touching the flowers and foliage after spraying. As it is the field bee workforce which usually comes in contact with the insecticides the best means of preventing accident lies in preventing that force from being affected. In the case of residual insecticides it is possible for some bees to return to the hive before dying and contaminate others, especially if they are collecting pollen and nectar from sprayed plants. If it is known that spraying is to be carried out near hives it is probably better to accept a day's loss of work rather than lose a lot of field bees. In this case the entrance to the hive can be closed with a fine-mesh wire and water can be provided by a feeder through a hole in the inner cover. A super can be placed over this and the cover put on top, as when feeding sugar syrup. Whenever possible nonresidual insecticides should be used in the vicinity of beehives so that the bees cannot be affected once the insecticide has dried out.

Wax pests

In addition to pests which attack the bees or honey, there are two wax pests which are important because they feed on wax. These are the caterpillars of two species of moth, the large wax moth *(Galleria mellonella)* and the small wax moth *(Achroia grisella)*. If the bee colony is vigorous the wax moth usually cannot gain a foothold but where the colony is weak the caterpillars may be able to eat away at the comb undisturbed. Infestation is started when the female moth lays eggs in the hive. The caterpillars which hatch feed on the wax, making webbing tunnels as they go. The tunnels and the fine black droppings of the caterpillar soon become obvious as the comb is gradually destroyed. Wax moths can very quickly find stored comb or any comb left lying about unprotected. For this reason it is essential that comb which is not in use be stored in sealed containers. Odd pieces of wax should never be left lying about and blocks of wax waiting to be sold should be kept in closed containers. When not in use, supers of frames can be stacked on a flat board and tablets of paradichlorbenzene placed between or on the frames and the whole stack covered with an inner cover and lid. Alternatively each super, with its frames, can be placed in a plastic bag (big plastic garbage bags or bags sold for lawn-clippings are useful) with a tablet or some flakes of paradichlorbenzene. It is essential to air the supers thoroughly for a day or two before bringing them back into service as the fumes of the fumigant are toxic to bees. Other fumigants are available, such as carbon bisulphide or methyl bromide, but these are more toxic to humans and domestic animals than paradichlorbenzene. Even this, however, should be handled only in a well-ventilated area, preferably in the open, because the fumes, although not unpleasant, can be toxic in the long term. Paradichlorbenzene kills larvae and moths.

CHAPTER NINETEEN
Bee Stings

When the word "bee" is mentioned most people think of one of two things—honey or stings—and most have an inherent fear of bees even if they have never been stung.

When a person is stung several things happen. There is the initial pain, which is not nearly as great as imagined. At the same time there is an element of surprise which is accompanied by at least some fright. This fright seems to be responsible for the exaggerated impression of pain because when beekeepers get over the feeling of fright the apparent pain is also not as severe. After experiencing a number of stings, provided there are no allergic or subsequent severe reactions, most beekeepers are not taken by surprise and lose their fear of stings.

When a person is stung the body sets in motion a number of complex reactions which eventually lead to the reduction of the effects of the sting so that no permanent harm results. There are, however, a few unfortunate individuals who are excessively allergic to bee stings and whose natural reactions result in undesirable effects.

After an initial feeling of pain, red swelling, varying in size according to individual reaction, occurs at the sting site and there may be a white area within the red swelling around the actual point of entry of the sting. Later the area may be itchy, and eventually the effects fade away. Many people react less and less to subsequent stings. By preparing oneself mentally for being stung, it is possible to suffer almost no feeling of pain or surprise at the sting.

In the case of an allergic person the effects of the stings usually become worse with each sting, and although the initial sting may have little effect, later stings may result in much more extensive swellings and other symptoms.

These symptoms also come on more rapidly. There may be a feeling of nausea and a general itching all over the body, accompanied by the appearance of red areas. Heartbeat rate increases and there is difficulty in breathing, even to the point of collapse. If there is any suggestion of symptoms other than those of local swelling medical advice should be sought immediately. Fortunately, few people are really allergic and one often hears of victims saying they are allergic when, in fact, they are referring to the normal local symptoms of a bee sting. Of course, even local reaction can be troublesome and even dangerous if the sting is in such places as on or near the eyes or throat. In general, take special care to avoid being stung anywhere about the neck and head.

Anyone who suspects that he is allergic should see a doctor before an emergency arises. Desensitising treatment is often possible and tablets for emergency treatment can usually be prescribed by a doctor. Allergic people should always have their tablets within reach.

When a bee stings the barbed end of the sting prevents its withdrawal. The glands which produce the venom and the venom storage organs are torn out of the bee's body. The bee eventually dies. As the venom continues to be injected for some time after insertion of the sting the effect can be lessened by its quick removal. If you grab the sting you are likely to squash the venom sac and inject more venom. It is better to remove it by scraping. Run your thumb nail across the skin to remove the sting. If you are wearing gloves the sharp edge of the hive tool is a good substitute.

When a bee stings it produces an odour which warns the other bees of danger and this tends to make them more inclined to sting. This is the reason for several bees stinging in quick succession during hive manipulation after a long period when none even attempted to sting.

When handling bees take precautions against being stung. There is nothing "brave" about laying yourself open to unnecessary stings. If you are stung, keep calm. The pain seems worse than it is because of the surprise. Remove the sting as quickly as possible to lessen the effects. If you react more violently then you should see a doctor immediately. Most people become less sensitive as they receive more stings until they reach the point at which they are hardly affected. In fact, many people believe that bee stings are instrumental in curing rheumatism and arthritis.

CHAPTER TWENTY
Legislation

Honey production is an important industry in Australia and it is essential that our bees be kept healthy and productive. For this reason each State has laws relating to the keeping of bees. These laws apply to everyone who has bees, whether it be one hive or a thousand.

In New South Wales each person owning bees must register with the Department of Agriculture. The registration fee depends on the number of hives owned but is, in any case, quite modest. If hives need to be destroyed to prevent disease spread it may be possible for the beekeeper to obtain compensation for loss. Each apiary must have at least one hive marked with the name and address or registration number of the owner so that it can be identified. Bees may only be kept in properly constructed hives so that the contents can be inspected easily. Whether it is in the comb or not, it is illegal to place honey (and wax) out in the open, where bees can gain access to it, because it can encourage robbing.

Diseased honey obtained in this way can be taken by the bees to healthy hives. It is essential to notify the Department of Agriculture or an Apiary Officer as soon as disease is found in any hives. Movement of bees from State to State is permitted only with the approval of appropriate State authorities and if you intend moving bees you should contact the Department of Agriculture to determine what is required. Imports into New South Wales require a health certificate from the exporting State.

The laws are aimed at preventing spread of diseases and it is in the beekeeper's interest to ensure that they are obeyed. It is his moral obligation to do everything possible to protect his own and other beekeepers' bees.

Some local Councils have regulations which restrict beekeeping in some ways, either to limited numbers of hives, or, where they have been ill advised, to certain times of year.

Glossary

Absconding   The whole colony leaves the hive and settles in another, usually more suitable place.

American brood disease   A bacterial disease of the brood.

Apiarist   Beekeeper.

Apiary   A group of hives arranged for ease of management at a suitable site.

Apiculture   The science and art of keeping bees.

Balling   Worker bees surrounding the queen in a mass with the intention of restraining or killing her.

Bee-bread   Pollen and honey mixture on which bees and brood feed.

Bee brush   A brush with soft bristles used to brush the bees from the comb.

Bee escape   A small piece of apparatus through which bees can pass in only one direction.

Bee space   The space between the various hive components within which bees will work with ease and not attempt to close up. An effective space is about 9–10 millimetres.

Beeswax   Wax secreted by worker bees from which they form the honeycomb.

Bottom board   The floor of the hive.

Brood   The young of the bee, whether eggs, larvae or pupae.

Brood box   The lowest hive body, or box, of the hive, in which the queen usually prefers to lay and which usually houses most of the brood.

Burr comb   Comb which the bees build in areas of the hive other than in the rectangle of the frames.

Capped brood   Pupae in cells which have been closed.

Capped honey   Cells containing ripe honey closed by wax for later use by bees.

Cappings   The remains of the caps of the cells after removal by the beekeeper prior to honey extraction.

Castes   The three kinds of bees in a colony—the queen, the drones and the workers each form a different caste.

Chunk honey   Pieces of comb immersed in bottled honey for sale.

Colony   The sum total of individual bees inhabiting a hive, normally consisting of a queen, drones, workers and brood.

Comb   Two layers of wax cells, arranged back to back, surrounded by the rectangular wooden frame in the hive.

Comb foundation   A manufactured, embossed wax sheet placed in a frame to guide comb development by the bees.

Corbicula   The structure on the hind leg of the worker bee in which the pollen is carried back to the hive.

Creamed honey   Honey which has been beaten so as to incorporate small bubbles in it.

Crop   Part of the bee's gut in which nectar is carried. Also called the honey sac.

Crystallised honey   Honey in which sugar crystals have precipitated from solution.

Cut-comb honey   Part of the comb which contains honey and has been cut out from the whole comb.

Dividing   Separating a colony into more than one part to increase the number of colonies.

Drone   The male bee whose sole purpose is to fertilise a virgin queen.

Embedding   Fixing comb foundation in a frame by means of reinforcing wires, usually using heat.

Embedding tool   A pair of small, toothed wheels on a handle used for attaching comb foundation to the frame wires.

European brood disease   A bacterial disease of brood.

Extracting The process of removing the honey harvest from the comb.

Extractor A machine in which frames are placed and spun around to remove honey from them.

Feeder A piece of equipment designed to allow bees access to a sweet liquid on which they feed.

Field bee An old bee, responsible for working away from the hive, collecting nectar, pollen, propolis and water.

Frame A rectangular wooden frame which can be removed from the hive and within the confines of which the bees make wax comb.

Green honey Nectar from which some water has been removed but which is not yet converted to "ripe" honey.

Hive Strictly, the structure used to house bees which constitute the colony. In Australia, the "hive" often refers to either structure or colony.

Hive body A box, part of the beehive, which contains frames.

Hive tool An all-purpose tool used when examining hives.

Honey sac See Crop.

House bee A newly emerged worker, responsible for domestic duties in the hive, such as ripening honey, producing wax, and guarding the hive.

Introducing cage A wire or wooden cage used to allow gradual introduction of a queen to a strange colony.

Nectar The sugary solution produced by plants usually, but not always, in the flowers.

Nectar flow A period when plants are producing large quantities of nectar which is available to the bees.

Nosema A disease of adult bees.

Nucleus A small colony in a small hive from which a full colony can be developed.

Nurse bee Young worker bee responsible for caring for the brood.

Overwintering The successful survival of winter by a hive.

Pollen Male reproductive material, produced by plants with a high protein content, used extensively by bees as food.

Pollen trap A piece of equipment which removes the pollen load from a worker bee as it enters the hive so that it can be collected.

Propolis Material of vegetable origin used by bees to seal and repair hives.

Queen Fertile female whose task is to lay enough eggs to maintain numbers in the colony.

Queen excluder A plate of metal, wire or plastic with holes or mesh big enough to allow workers but not queen bees to pass through.

Rabbet A metal insert inside the ends of the hive body on which the frames are suspended.

Requeening The introduction of a new queen to replace an old or abnormal one.

Robbing Stealing honey from one colony of bees by another. This term should not be used for harvesting the crop by the beekeeper.

Royal jelly Substance produced by the workers and fed to all young larvae, older larvae destined to become queens, and the queen herself.

Sac brood A virus disease affecting brood and adult bees.

Section honey Honey in comb, produced in small rectangular wooden frames for sale as such, and not extracted.

Super A hive body placed above the brood box in which honey is stored by the bees. Sometimes the brood is extended into a super.

Supersedure The process whereby the colony replaces a queen without swarming.

Swarm The queen, some of the workers, and drones from a colony which leave to found a new one elsewhere.

Tarsus The small segments of the bee's leg farthest from the body.

Uncapping Removal of cappings from the comb prior to extraction.

Wax embedder A piece of mechanical or electrical apparatus for embedding the reinforcing wires in a sheet of comb foundation.

Wax moth Moth pests of wax, the caterpillar stages of which eat the wax.

Worker The smallest and most numerous bee caste, the worker is an infertile female and is responsible for most of the day to day work in the colony.

Further Reading

The Amateur Beekeeper. A monthly newsletter issued to members. Amateur Beekeepers' Association.

Clemson, Alan. *Honey and Pollen Flora.* New South Wales Department of Agriculture, Sydney, 1985.

Bees and Honey (several editions). New South Wales Department of Agriculture, Sydney.

"Beginning in Bees". *Agfact* Pamphlet A8.9.1. New South Wales Department of Agriculture, Sydney, 1981.

Redpath, Norman. *A Guide to Keeping Bees in Australia.* Nelson, Melbourne, 1980.

Index

Index of honey and pollen flora